Life in Extreme Environments - Diversity, Adaptability and Valuable Resources of Bioactive Molecules

Edited by Afef Najjari

Published in London, United Kingdom

Life in Extreme Environments - Diversity, Adaptability and Valuable Resources of Bioactive Molecules
http://dx.doi.org/10.5772/intechopen.100892
Edited by Afef Najjari

Contributors
Hulya Celik, Kader İlhan, Ahmed M. Mohammed Shuikan, Ibrahim A. Arif, Rakan M. Alshuwaykan, Nawab Ali, Muhammad Nughman, Syed Majid Shah, Imene Rajhi, Bechir Baccouri, Safa Khalifa, Fethi Barhoumi, Moez Amri, Haythem Mhadhbi, Afef Najjari

First published in London, United Kingdom, 2023 by IntechOpen
IntechOpen is the global imprint of INTECHOPEN LIMITED, registered in England and Wales, registration number: 11086078, 5 Princes Gate Court, London, SW7 2QJ, United Kingdom

British Library Cataloguing-in-Publication Data
A catalogue record for this book is available from the British Library

Additional hard and PDF copies can be obtained from orders@intechopen.com

Life in Extreme Environments - Diversity, Adaptability and Valuable Resources of Bioactive Molecules
Edited by Afef Najjari
p. cm.
Print ISBN 978-1-80356-818-8
Online ISBN 978-1-80356-819-5
eBook (PDF) ISBN 978-1-80356-820-1

Meet the editor

Afef Najjari is an assistant professor of bioinformatics at the Faculty of Sciences, University of Tunis El Manar in Tunisia. Dr. Najjari has worked on a variety of national and international projects related to genomic and enzymatic diversity in lactic acid bacteria, as well as the diversity and adaptation of extremophilic microbes. Her current research interests include metagenomic data analysis, genome assemblies and annotations, and the development of biological databases. Dr. Najjari has published several articles in highly ranked journals and has co-edited several books. She has also contributed to building capacity in bioinformatics at national and international levels.

Contents

Preface

Extremophiles are a group of microbes and organisms that possess the ability to live in extreme physico-chemical and nutritional environments. They include acidophiles, alkalophiles, halophiles, barophiles, (hyper) thermophiles, and psychrophiles. These organisms have evolved several mechanisms to ensure genomic integrity, cell division, and energy conservation in extreme conditions. Recently, researchers have begun to study the phylogenetic relationship between extremophilic microbes through the analysis of their genome sequences, enabling identification of distinctive genes and metabolic pathways involved in the extremophilic way of life. These analyses have provided key data on the functionality of genes across organisms and environments, helping us to track the evolutionary events involved in environmental adaptations at the population and strain level. The six chapters of this book are: (1) Introductory Chapter: Brief Overview of the Diversity of Secondary Metabolite in Extreme Environment – The Case of Halobacteria Based on Genome Sequence Mining; (2) The Role of Metagenomic Approaches in the Analysis of Microbial Community in Extreme Environment; (3) Genotype-specific Patterns of Physiological, Photosynthetic, and Biochemical Responses in Faba Bean Contrasting Pair to Salinity; (4) Antioxidant Effective Aromatic Compounds; (5) Extremophiles and Limits of Life in a Cosmic Perspective; and (6) Genome Analysis Provides Insights into the Osmoadaptation Mechanisms of *Halomonas titanicae*.

Afef Najjari
Faculty of Sciences of Tunisia,
University of Tunis El Manar,
Tunis, Tunisia

Chapter 1

Introductory Chapter: Brief Overview of the Diversity of Secondary Metabolite in Extreme Environment – The Case of Halobacteria Based on Genome Sequence Mining

Afef Najjari

1. Introduction

Secondary metabolites (SMs) are a range of bioactive compounds yielded by several microorganisms, such as bacteria, fungi, and archaea [1]. They are not directly involved in basic life processes, such as growth, cell division, and respiration [2]. However, they are thought to be important components of the innate immune system against other organisms and also play essential roles in enhancing tolerance to environmental stress for some microbes, such as Haloarchaea domain [3]. Members of Halorarchaea are reported to generate various types of molecules, such as terpenes, polyketides, alkaloids, and archaeocins, which can be exploited for biotechnological purposes [1]. *Halobacteria* constitute an evolutionary distinct salt-tolerant microorganism and are known as Haloarchaea that require a minimum of 8% salt concentration to grow [4]. Actually, Haloarchaea is divided into three orders: Halobacteriales, Haloferacales, and Natrialbales consisting currently of 48 genera [5]. It is worth noting that compared with bacteria or plants, SMs in Haloarchaea have been far less researched. Here, we aimed to predict the SMs of genomes sequences of 48 genera genus available in the IMG database [5] using antiSMASH version 6.0 with default parameters [6]. Indeed, computation of strain similarities based on SMs production profile was conducted with MVSP (Multi-Variate Statistical Package) software. The similarity of SMs profiles among genera was calculated using Pearson's product correlation coefficient. The clustering of strains was based on the unweighted pair group method with an arithmetic average [7].

2. Results

Results showed that nine (n = 9) SBs were identified within 46 (of 48) genome sequences (**Figure 1**):

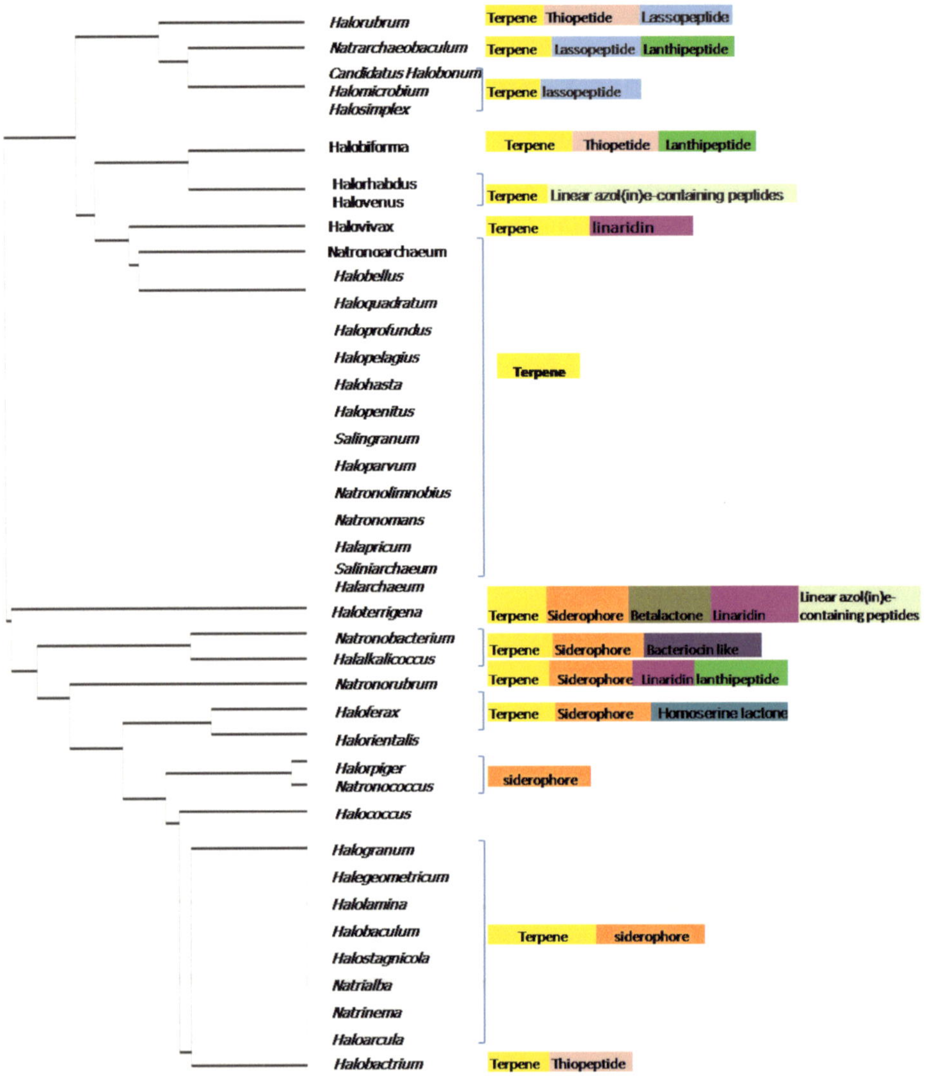

Figure 1.
Unweighted pair group method with arithmetic mean (UPGMA) dendrogram derived from similarity coefficients calculated by Pearson's product moment, showing the relationship among Haloarchaeal genera analyzed by presence or absence of secondary metabolites types. The different color refers to the nine secondary metabolites identified.

a. **Terpenes** are a major class of biological compounds usually found in the plant known to protect the plant against and in bacteria [8]. Here, the analysis showed that almost all genomes could produce terpenes (**Figure 1**);

b. **Siderophore** are organic compounds with low molecular masses that are produced by microorganisms and plants growing under low iron conditions [9, 10]. There are only very few studies on siderophores reported on Haloarchaea [9] Here, the analysis showed that 18 genera could produce siderophores (**Figure 1**);

c. **Ribosomally synthesized and posttranslational modification modified peptides with** antibiotic potential found in Archaea [11], including

(i) Lantipeptide [12], here, genomes mining showed that *Natrarchaeobaculum*, *Natronorubrum,* and *Halobiforma* could produce Lantipeptides **(ii)** Thiopetide [13] identified in three genera *Halorubrum*, *Halobiforma,* and *Halobacterium* **(ii)** Lasso peptide, recently detected across some archaeal genomes [14], here identified in *Halorubrum*, *Halosimplex*, *Halomicrobium*, *Natrarchaeobaculum,* and *Candidatus Halobonum* **(iv)** Linaridin with antibiotic potential [15] was found within *Natronoarchaeum*, *Haloterrigena,* and *Natronorubrum* **(v)** Linear azol(in) e-containing peptides [16] identified in *Haloterrigena* and *Halovivax*.

d. The remaining SMs, including acyl homoserine lactones (AHLs) [17], metabolites controlling a range of quorum sensing phenotypes in bacteria was identified in Halorientalis and Haloferax. The betalactone [18] was identified only in two genera, *Halococcus* and *Haloterrigena*.

On the basis of our *in silico* analyses, we can conclude that *Haloterregina* represents a good candidate for SMs production profile (terpene, siderophore, betalactone, linaridin, and Linear azol(in)e-containing peptides), followed by *Natronorubrum* (terpene, siderophore, betalactone, and Lantipeptide). This chapter will open new research lines that will shed light on metabolites in extreme environments and their biotechnological potential.

Author details

Afef Najjari
Faculty of Sciences of Tunisia, University of Tunis El Manar, Tunis, Tunisia

*Address all correspondence to: afef.najjari@fst.utm.tn; najjariafef@gmail.com

References

[1] Corral P, Amoozegar MA, Ventosa A. Halophiles and their biomolecules: Recent advances and future applications in biomedicine. Marine Drugs. 2019;**18**(1):33. DOI: 10.3390/md18010033

[2] Piasecka A, Jedrzejczak-Rey N, Bednarek P. Secondary metabolites in plant innate immunity: Conserved function of divergent chemicals. The New Phytologist. 2015;**206**:948-964. DOI: 10.1111/ nph.13325

[3] Finking R, Marahiel MA. Biosynthesis of nonribosomal peptides. Annals of Review Microbiology. 2004;**58**:453-488. DOI: 10.1146/annurev. micro.58.030603.123615

[4] Oren A. Microbial life at high salt concentrations: Phylogenetic and metabolic diversity. Saline System. 2008;**4**:2

[5] Parte AC, Sardà CJ, Meier-Kolthoff JP, Reimer LC, Göker M. List of prokaryotic names with standing in nomenclature (LPSN) moves to the DSMZ. International Journal of Systematic and Evolutionary Microbiology. 2020;**70**:5607-5612. DOI: 10.1099/ ijsem.0.004332

[6] Blin K, Shaw S, Kloosterman AM, Charlop-Powers Z, van Wezel GP, Medema MH, et al. antiSMASH 6.0: improving cluster detection and comparison capabilities. Nucleic Acids Research. 2021;**49**:W29-W35. DOI: 10.1093/nar/gkab335

[7] Sneath PHA, Sokal RR. Numerical Taxonomy. San Francisco: W.H. Freeman and company; 1973

[8] Dickschat JS, Rabe P, Citron CA. The chemical biology of dimethylsulfoniopropionate. Organic & Biomolecular Chemistry 2015 Feb 21;13(7):1954-1968. doi: 10.1039/ c4ob02407a. PMID: 25553590.

[9] Dave BP, Anshuman K, Hajela P. Siderophores of halophilic archaea and their chemical characterization. Indian Journal of Experimental Biology. 2006;**44**(4):340-344

[10] Niessen N, Soppa J. Regulated iron siderophore production of the halophilic archaeon *Haloferax volcanii*. Biomolecules. 2020;**10**(7):1072. DOI: 10.3390/biom10071072

[11] Jančič S, Frisvad JC, Kocev D, Gostinčar C, Džeroski S, Gunde-Cimerman N. Production of secondary metabolites in extreme environments: Food- and airborne Wallemia spp. produce toxic metabolites at hypersaline conditions. PLoS One. 2016;**11**(12):e0169116. DOI: 10.1371/ journal.pone.0169116

[12] Skinnider MA, Johnston CW, Edgar RE, et al. Genomic charting of ribosomally synthesized natural product chemical space facilitates targeted mining. Proceedings of the National Academy Science. 2016;**113**:E6343-E6351. DOI: 10.1073/pnas.1609014113

[13] Schwalen CJ, Hudson GA, Kille B, Mitchell DA. Bioinformatic expansion and discovery of thiopeptide antibiotics. Journal of American Chemical Society. 2018;**140**(30):9494-9501. DOI: 10.1021/ jacs.8b03896

[14] Maksimov MO, Pelczer I, Link AJ. Precursor centric genome-mining approach for lasso peptide discovery. Proceedings of the National Academy Science USA. 2012;**109**:15223-15228. DOI: 10.1073/pnas1208978109

[15] Claesen J, Bibb M. Genome mining and genetic analysis of cypemycin biosynthesis reveal an unusual class of posttranslationally modified peptides. Proceedings of the National Academy Science USA. 2010;**107**(37):16297-16302. DOI: 10.1073/pnas.1008608107

[16] Cox CL, Doroghazi JR, Mitchell DA. The genomic landscape of ribosomal peptides containing thiazole and oxazole heterocycles. BMC Genomics. 2015;**16**:778. DOI: 10.1186/s12864-015-2008-0

[17] Charlesworth J, Kimyon O, Manefield M, Beloe CJ, Burns BP. Archaea join the conversation: detection of AHL-like activity across a range of archaeal isolates. FEMS Microbiology Letters. 2020;**367**(16):fnaa123. DOI: 10.1093/femsle/fnaa123

[18] Wackett LP. Microbial β-lactone natural products: An annotated selection of World Wide Web sites relevant to the topics in microbial biotechnology. Microbes Biotechnology. 2016;**10**(1):218-220. DOI: 10.1111/1751-7915.12600

Chapter 2

The Role of Metagenomic Approaches in the Analysis of Microbial Community in Extreme Environment

Ahmed M. Shuikan, Rakan M. Alshuwaykan and Ibrahim A. Arif

Abstract

Metagenomic is a promising technique that has many applications in different fields. In fact, metagenomics is the ideal culture-independent technique that unravels the microbial composition and biodiversity in the sample, which helps scientists to study and understand how this biodiversity is affected by continuously changing conditions in the environment and how this microbial community interacts with each other. In the past, the microbial composition in extreme environments was undiscovered due to the difficulty of isolation, culturing, and identification of microbes living there. However, nowadays after the development and combination of metagenomic and next-generation sequencing techniques, it became more easy to study the microbial composition in extreme environments without culturing. In this chapter, the use of metagenomic techniques to study the microbial biodiversity in different extreme environments are discussed. In addition, different NGS platforms are discussed in terms of principles, advantages, and limitations.

Keywords: metagenomics, extreme environment, salinity soil, microbial taxonomy, biodiversity, sequencing approaches

1. Introduction

Microorganisms are crucial to the control of many activities, including the recycling of essential elements and nutrients, the dynamics of the biogeochemical cycles (such as carbon, nitrogen, and oxygen cycles), and the development of soil structure. To increase our knowledge of microbial diversity, activity, and interactions with different ecosystem components, it is essential to have an understanding of the structure, function, and activities of microbes [1]. Prokaryotic taxonomy is the term used to describe the traditional classification of the diversity of microorganisms based on the ideas of classification, naming, and characterization. Based on their morphology (form), growth environments, bacterial pathogenicity, and early investigations introduced the ideas of genera and species into the taxonomy of bacteria [2].

Later, the classification scheme developed by the American Society for Microbiology included the biochemical and physiological characteristics of bacteria. DNA–DNA hybridization (DDH) methods have been used as the de facto method for prokaryotic classification based on genomic similarity since the 1960s [3].

As a way to distinguish between different bacterial species and research bacterial phylogeny, 16S rRNA sequence similarity emerged as one of the most used methods [4]. Later, following the discovery of the polymerase chain reaction (PCR), molecular methods to study prokaryotic classification were developed, including denaturing gradient gel electrophoresis (DGGE), temperature gradient gel electrophoresis (TGGE), restriction fragment length polymorphism (RFLP), terminal restriction fragment length polymorphism (TRFLP), ribosomal intergenic spacer analysis (RISA), and real-time PCR (quantitative PCR) [5]. The field of microbial taxonomy has been transformed most significantly by high-throughput DNA sequencing technologies, such as metagenomics shotgun sequencing.

When researching soil microbiology, genomics is especially crucial. According to different researches, one gram of soil may contain more different microorganisms than that have been grown in laboratory to date [6]. Therefore, metagenomics appears to be the best culture-independent method for revealing soil biodiversity and researching how this biodiversity is impacted by constantly changing environmental factors. The first step in the approach for metagenomics research is to choose an appropriate ecological or biological setting that supports a diverse range of microbial communities with potential biotechnological and therapeutic uses. Extreme conditions are mostly what draw metagenomic researchers to certain environments. These include highly alkaline or acidic pH conditions, high metal concentrations, pressures, or radiation, as well as settings with high salinity or extreme temperatures [7].

Beginning with the isolation of genomic DNA from the soil sample that represents the entire population, a DNA library is built from the isolated DNA, and the DNA library is then screened for a target sequence (**Figure 1**). It is crucial to choose a DNA extraction procedure that will generate enough DNA. The diversity of the entire microbial community in the target environment. One of the most difficult steps in the metagenomic analysis is still this one. Depending on the type of soil analyzed, the chemical and physical features of soils are quite diverse and complex, which will make it challenging to design a reference method for DNA extraction from soils. Additionally, a variety of chemicals found in soils co-extract genomic DNA and have an inhibiting influence on the subsequent processing of extracted DNA. Humic and fulvic acids are two examples [8]. As a result, for each kind of soil, optimization and comparison of various extraction methods are typically necessary [9–11].

The genomic DNA that was extracted from the target environment is then used to create a DNA library. This is accomplished by first dividing the isolated DNA into pieces of the proper sizes to facilitate cloning. Either mechanical shearing or restriction enzyme digestion is used to accomplish this. These procedures produce fragments of DNA that are then cloned into the appropriate cloning vector. Short-insert genomic libraries are created by inserting small DNA fragments into plasmid vectors. Large inserts can be cloned onto BAC or cosmid vectors, which can carry inserts larger than 40 Kb, or fosmid or cosmid vectors, which can handle inserts up to 40 Kb in size [12].

1.1 Metagenomic methods

The term "metagenomics" refers to a collection of methods and techniques used to analyze the entire genomes of microorganisms residing in a given environment

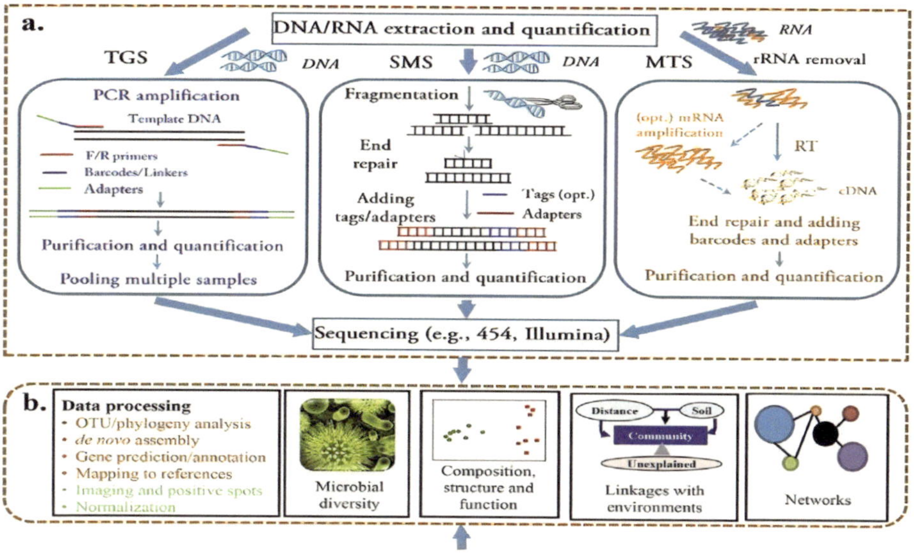

Figure 1.
Metagenomic protocol. (a) the first step of metagenomic analysis is DNA extraction by using different DNA extraction techniques and (b) the second step of metagenomic analysis by using software data analysis to analyze the huge data collected from samples.

without regard to culture [13]. It has a wide range of beneficial applications and shows great promise in both medical and environmental microbiology. Metagenomics is most frequently used in environmental microbiology to research the total genomic DNA analysis that has been used to study the diversity of microbial communities in various settings, how various microorganisms interact with one another there, and how these communities adapt to changes in the physical and chemical characteristics of these environments [14]. Additionally, metagenomics offers the chance to find and isolate new enzymes with industrial uses from harsh habitats where uncultivable extremophiles reside. Functional metagenomics enables the isolation of genes encoding extremozymes, enzymes capable of catalyzing in severe settings, or genes that will improve understanding of the mechanisms that make such organisms resistant to extreme environmental conditions in such situations [7].

E.coli, a microbe that has been extensively studied and is simple to work with in the lab, is typically used to create DNA libraries. Shuttle vectors are employed to transfer the libraries into the appropriate host in the event that the genes contained in DNA inserts need to be expressed in other microorganisms [15].

Finally, a screening test is used to look for a gene that performs a certain function, and the gene product is then functionally examined. There are two distinct metagenomics approaches that are frequently applied in studies. The first one focuses on using marker genes, such as the ribosomal genes 16S rRNA [16] and 18S rRNA, to research the makeup of microbial communities in certain environments or a specific protein-coding gene of medical or industrial value [17, 18]. Targeted metagenomics is the name given to this tactic. Shotgun metagenomics is the second method. In this method, high-throughput next-generation sequencing is used to get broad coverage of genomic DNA sequences in order to evaluate the overall taxonomic structure or functional potential of microbial communities [19].

Recent times have seen a significant increase in the use of various next-generation sequencing (NGS) platforms, each with its own advantages and disadvantages, for the taxonomic profiling, characterization, and analysis of microbial communities. High-throughput, short read sequences, and relatively declining costs are all characteristics of metagenomic samples. These platforms are beneficial in that they do not require DNA fragment cloning [20]. Recent improvements in NGS technology have been made to accommodate a wide range of applications, costs, and capabilities [21]. The 454 life sciences (Roche) and Illumina systems (Solexa) platforms are the most often utilized ones [22]. The 454-sequencing technology, which was the first next-generation technology to be commercially accessible, is based on the pyrosequencing method. It offers analysis with a high-throughput at a reasonable cost [23]. This method sequencing procedure involves the insertion of nucleotides into the capture of the released pyrophosphate, which undergoes an enzyme process to produce light, and allows for the detection of the developing chain. In order to assign a distinct nucleotide to each nucleotide incorporation event, different nucleotides are successively added. The light signals are finally transformed into sequencing data. The DNA fragments are fixed on beads in a water–oil emulsion before being amplified in a 454 pyrosequencer [24]. The thermophilic cellulose-degrading microbial species isolated from the hot springs in Xiamen, China, have been studied using pyrosequencing. It was also widely used to analyze the diversity of microbes in a variety of habitats, such as different soil environments [25–28] and marine environments.

Reversible terminator nucleotides that are fluorescently tagged are a key component of Illumina sequencing technology. The terminator nucleotides are coupled with blocking groups that may be removed from the nitrogen base in a single step, unlike deoxynucleotides, which are chemically changed to stop further DNA synthesis as is the case with sanger sequencing. On a chip with attached primers, DNA synthesis takes to happen. Following each cycle of synthesis, a laser is used to excite the dyes attached to each nucleotide. This is followed by scanning of the integrated bases. The blocking group and the dye must first be eliminated by a chemical reaction in order for the subsequent synthesis cycle to start. Multiple settings, including freshwater sponges, the gastrointestinal system, soils, and marine environments, were effectively studied using the Illumina sequencing technology.

In addition to the methods already described, metagenomics research also makes use of recently developed sequencing technology. These include the single-molecule real-time (SMRT) DNA sequencing from Pacific Biosciences, Ion Torrent's semiconductor sequencing, and Applied Biosystems' SOLiD 5500 W Series [22]. Technologies that are more advanced and cutting-edge are being created, and they could be very helpful in metagenomics research. Oxford nanopore technologies are actively working on strand sequencing technologies, which make it possible to sequence whole DNA strands as they travel through a protein nanopore [29]. One of the most exciting new technologies in the genomics era is Irys technology, created by BioNano genomics. The analysis of the enormous number of sequence data that are produced from the screening phase in metagenomics is the component of the procedure that is the most difficult built-in library many bioinformatics tools have grown over time to assist in analyzing the metagenomic data and comparing it to online databases.

1.2 Metagenomics to identify microbial diversity in extreme environments

Both industrial microbiologists and ecologists are interested in the study of microbial communities in extreme conditions, such as saline soil or other saline habitats, such as saline waterways or saline sediments [30]. Understanding the impacts of salinity on soil ecosystems can aid in understanding how the structure of the microbial community changes in response to changes in salinity. However, due to their benefits over mesophilic enzymes in commercial applications, industrial microbiologists are more interested in isolating halophilic bacteria or organisms that like to survive in salty environments, or their enzymes.

Vera-Gargallo [31] used metagenomics to explore the microbial communities in two hypersaline soils in the Odiel saltmarshes of Spain. They then matched their findings to the information found in the databases for the 13 saline aquatic habitats. In contrast to hypersaline aquatic habitats, they discovered that hypersaline soils had. more diversified microbial communities that include non-halophilic organisms. In order to understand how salt impacts microbial dispersion in saline environments, more research has recently been focused on saline soils. Using high-throughput sequencing technology, it was recently shown that bacterial communities were more abundant in surface soils (0–10 cm) than in deep soils in the saline soils of Qarhan Sale Lake, China (15–30 cm). This was linked to the fact that underground soils had low oxygen content in other investigations as well [32]. *Proteobacteria, Bacteroidetes,* and *Gemmatimonadetes* were discovered to be the three phyla with the greatest abundance in this study. *Proteobacteria* contain the *Alphaproteobacteria*, *Deltaproteobacteria,* and *Betaproteobacteria* and were the next most prevalent classes after *Gammaproteobacteria* [33]. Previous research [34] revealed similar findings. Nevertheless, there was no discernible variation in the microbial community between the surface and deep saline soils.

The structure and metabolic processes of microbial communities are being impacted by soil salinity, according to more recent studies. According to a study by Chen [35], the phyla *Planctomycetes* and *Bacteroidetes* of bacteria were observed to decline under continuous irrigation with saline water, while *proteobacteria, Actinobacteria,* and *Chloroflexi* grew. They also discovered that irrigation with salt water has a significant impact on the metabolic activities of soil microorganisms. The salinity of irrigation was observed to boost soil bacterial richness.

Uncertainty surrounds the exact method by which salinity alters the composition of microbial communities. Morrissey [36] hypothesized that microbes' preferences for saline soils are influenced by their evolutionary history. They discovered that most *proteobacteria* favored saltwater, while many *proteobacteria* chose freshwater in their study of wetland soils. There was a connection between the quantity of microorganisms that responded to salinity and phylogenetically grouped salinity preferences. Analysis of the 16S rRNA gene sequences of bacteria isolated from soil in Italy's A horizon that has varied salt concentrations. The findings demonstrated that variations in soil salt concentration cause differences in the composition of bacterial communities. *Proteobacteria* > *Actinobacteria* > *Acidobacteria* > *Verrucomicrobia* > *Gemmatimono dates* > *Firmicutes* > *Chloroflexi* > *Bacteriodes* > *Chlorobi* were the most prevalent bacterial species in this soil in terms of abundance. However, they discovered that various levels of some taxonomic groups were present in the soil that had varying salt values. Nevertheless, several bacterial species did not have salt levels, which were present in equal amounts at all of the investigated sites [37].

The same bacterial groups that were found in the prior research were also found in large numbers in saline soils [38]. Functional genomics has been effectively used to separate the genetic components of osmoadaptation from the isolated metagenome library. In order to better understand the mechanisms by which the microbial population responds to salinity, they were able to locate salt-tolerant genes. Through genetic engineering, numerous genes have been cloned to enhance salt tolerance in crops, such as *P5CS, DREB1A,* and *AtNHX1.*

By using metagenomic analysis, several more severe settings have been examined. Organic carbon sources are scarce in alkaline environments, such as soda lakes and low-saline-alkaline habitats. Understanding the diversity of microbes present in these harsh conditions, according to scientists, may help us better comprehend the early and rare forms of life [39]. Five novel species of the candidate phyla radiation were abundantly found in the hypersaline soda lake, according to a recent metagenomic investigation [40]. To investigate the microbial diversity in contaminated soils, metagenomics is a potent investigative tool. Heavy metal contamination of the soil at mining sites can be detrimental. The most prevalent bacteria in these conditions were determined to be *Solirubrobacter*, *Geobacter, Edaphobacter, and Pseudomonas* [41]. Other environments that metagenomics is probably being used to investigate acidic, low-oxygen, and volcanic settings are among the methods [42–44].

2. Conclusion

Due to the presence of non-cultivable microorganisms, traditional culturing techniques, such as isolation, do not reflect the real structure of microbial community when studying extreme environments. However, the emergence and development of metagenomic techniques boosted the identification and profiling of the microbial community in extreme environments and allowed to overcome non-cultivable microorganisms issues. Nowadays, by using metagenomic approaches scientists are able to profile microbial community and investigate the biodiversity in a given environment and also able to study and understand how these microorganisms have adapted to such environments. Resulting in, many different secondary metabolites have been discovered that are produced by microorganisms in order to adapt to extreme environments. The 16S rRNA gene is the most suitable gene to study microbial biodiversity and establish a phylogenetic relationship among unculturable and novel microorganisms.

Author details

Ahmed M. Shuikan[1*], Rakan M. Alshuwaykan[2] and Ibrahim A. Arif[1]

1 Botany and Microbiology Department, College of Science, King Saud University, Riyadh, Saudi Arabia

2 Suergery Department, King Khalid Civilian Hospital, Tabuk, Saudi Arabia

*Address all correspondence to: ahmedshuikan@gmail.com

References

[1] Quince C, Curtis TP, Sloan WT. The rational exploration of microbial diversity. The ISME Journal. 2008;**2**(10):997

[2] Drews G. The roots of microbiology and the influence of Ferdinand Cohn on microbiology of the 19th century. FEMS Microbiology Reviews. 2000;**24**(3):225-249

[3] Grimont PA. Use of DNA reassociation in bacterial classification. Canadian Journal of Microbiology. 1988;**34**(4):541-546

[4] Lane DJ, Pace B, Olsen GJ, Stahl DA, Sogin ML, Pace NR. Rapid determination of 16S ribosomal RNA sequences for phylogenetic analyses. Proceedings of the National Academy of Sciences. 1985;**82**(20):6955-6959

[5] Stefanis C, Alexopoulos A, Voidarou C, Vavias S, E. & Bezirtzoglou. Principal methods for isolation and identification of soil microbial communities. Folia Microbiologia (Praha). 2013;**58**(1):61-68

[6] Torsvik V, Øvreås L. Microbial diversity and function in soil: From genes to ecosystems. Current Opinion in Microbiology. 2002;**5**(3):240-245

[7] Mirete S, Morgante V, González JE. Functional metagenomics of extreme environments. Current Opinion in Biotechnology. 2016;**38**:143-149

[8] Young JM, Rawlence NJ, Weyrich LS, Cooper A. Limitations and recommendations for successful DNA extraction from forensic soil samples: A review. Science & Justice. 2014;**54**(3):238-244

[9] Finley SJ, Lorenco N, Mulle J, Robertson BK, Javan GT. Assessment of microbial DNA extraction methods of cadaver soil samples for criminal investigations. Australian Journal of Forensic Sciences. 2016;**48**(3):265-272

[10] Lim NY, Roco CA, Frostegård Å. Transparent DNA/RNA Coextraction workflow protocol suitable for inhibitor-rich environmental samples that focuses on complete DNA removal for Transcriptomic analyses. Frontiers in Microbiology. 2016;**7**:1588

[11] Mazziotti M, Henry S, Laval P, Bonnefoy A, Falla J. Comparison of two bacterial DNA extraction methods from non-polluted and polluted soils. Folia Microbiologica. 2018;**63**(1):85-92

[12] Simon C, Daniel R. Construction of small-insert and large-insert metagenomic libraries. In: Metagenomics. New York, NY: Humana Press; 2017. pp. 1-12

[13] Daniel R. The metagenomics of soil. Nature Reviews Microbiology. 2005;**3**(6):470

[14] Delmont TO, Robe P, Cecillon S, Clark IM, Constancias F, Simonet P, et al. Accessing the soil metagenome for studies of microbial diversity. Applied and Environmental Microbiology. 2011;**77**(4):1315-1324

[15] Lam KN, Cheng J, Engel K, Neufeld JD, Charles TC. Current and future resources for functional metagenomics. Frontiers in Microbiology. 2015;**6**:1196

[16] Païssé S, Valle C, Servant F, Courtney M, Burcelin R, Amar J, et al. Comprehensive description of blood microbiome from healthy donors assessed by 16 S targeted

metagenomic sequencing. Transfusion. 2016;**56**(5):1138-1147

[17] Nurdiani D, Ito M, Maruyama T, Terahara T, Mori T, Ugawa S, et al. Analysis of bacterial xylose isomerase gene diversity using gene- targeted metagenomics. Journal of Bioscience and Bioengineering. 2015;**120**(2):174-180

[18] Lanza VF, Baquero F, Martínez JL, Ramos R, Gonzalez B, Andremont A, et al. In-depth resistome analysis by targeted metagenomics. Microbiome. 2018;**6**(1):11

[19] Quince C, Walker AW, Simpson JT, Loman NJ, Segata N. Shotgun metagenomics, from sampling to analysis. Nature Biotechnology. 2017;**35**(9):833

[20] Ansorge WJ. Next-generation DNA sequencing techniques. New Biotechnology. 2009;**25**(4):195-203

[21] Scholz L, Chain PS. Next generation sequencing and bioinformatic bottlenecks: The current state of metagenomic data analysis. Current Opinion in Biotechnology. 2012;**23**(1):9-15

[22] Oulas A, Pavloudi C, Polymenakou P, Pavlopoulos GA, Papanikolaou N, Kotoulas G, et al. Metagenomics: Tools and Insights for Analyzing next-generation sequencing data derived from biodiversity studies. Bioinformatics and Biology Insights. 2015;**9**:BBI-S12462

[23] Ronaghi M. Pyrosequencing sheds light on DNA sequencing. Genome Research. 2001;**11**(1):3-11

[24] Goodwin S, McPherson JD, McCombie WR. Coming of age: Ten years of next-generation sequencing technologies. Nature Reviews Genetics. 2016;**17**(6):333

[25] Egge E, Bittner L, Andersen T, Audic S, de Vargas C, Edvardsen B. 454 pyrosequencing to describe microbial eukaryotic community composition, diversity and relative abundance: A test for marine haptophytes. PLoS One. 2013;**8**(9):e74371

[26] Fan W, Huo G, Li X, Yang L, Duan C, Wang T, et al. Diversity of the intestinal microbiota in different patterns of feeding infants by Illumina high-throughput sequencing. World Journal of Microbiology and Biotechnology. 2013;**29**(12):2365-2372

[27] Hong C, Si Y, Xing Y, Li Y. Illumina MiSeq sequencing investigation on the contrasting soil bacterial community structures in different iron mining areas. Environmental Science and Pollution Research. 2015;**22**(14):10788-10799

[28] Lentini V, Gugliandolo C, Bunk B, Overmann J, Maugeri TL. Diversity of prokaryotic community at a shallow marine hydrothermal site elucidated by Illumina sequencing technology. Current Microbiology. 2014;**69**(4):457-466

[29] Jain M, Olsen HE, Paten B, Akeson M. The Oxford Nanopore MinION: Delivery of nanopore sequencing to the genomics community. Genome Biology. 2016;**17**(1):239

[30] Shuikan AM, Hozzein WN, Alzharani MM, Sandouka MN, Al Yousef SA, Alharbi SA, et al. Enhancement and identification of microbial secondary metabolites. In: Extremophilic Microbes and Metabolites - Diversity, Bioprespecting and Biotechnological Applications. London: IntechOpen; 2020

[31] Vera-Gargallo B, Ventosa A. Metagenomic insights into the phylogenetic and metabolic diversity of the prokaryotic community dwelling in Hypersaline soils from the Odiel saltmarshes (SW Spain). Genes. 2018;**9**(3):152

[32] Serkebaeva YM, Kim Y, Liesack W, Dedysh SN. Pyrosequencingbased assessment of the bacteria diversity in surface and subsurface peat layers of a northern wetland, with focus on poorly studied phyla and candidate divisions. PLoS One. 2013;**8**(5):e63994

[33] Xie K, Deng Y, Zhang S, Zhang W, Liu J, Xie Y, et al. Prokaryotic community distribution along an ecological gradient of salinity in surface and subsurface saline soils. Scientific Reports. 2017;**7**(1):13332

[34] Ma B, Gong J. A meta-analysis of the publicly available bacterial and archaeal sequence diversity in saline soils. World Journal of Microbiology and Biotechnology. 2013;**29**(12):2325-2334

[35] Chen L, Li C, Feng Q, Wei Y, Zheng H, Zhao Y, et al. Shifts in soil microbial metabolic activities and community structures along a salinity gradient of irrigation water in a typical arid region of China. Science of the Total Environment. 2017;**598**:64-70

[36] Morrissey EM, Franklin RB. Evolutionary history influences the salinity preference of bacterial taxa in wetland soils. Frontiers in Microbiology. 2015;**6**:1013

[37] Canfora L, Bacci G, Pinzari F, Papa GL, Dazzi C, Benedetti A. Salinity and bacterial diversity: To what extent does the concentration of salt affect the bacterial community in a saline soil? PLoS One. 2014;**9**(9):e106662

[38] Ahmed V, Verma MK, Gupta S, Mandhan V, Chauhan NS. Metagenomic profiling of soil microbes to mine salt stress tolerance genes. Frontiers in Microbiology. 2018;**9**:159

[39] Veríssimo A, Tiago I. Bacterial diversity in a nonsaline alkaline environment. In: Nelson KE, editor. Encyclopedia of Metagenomics: Environmental Metagenomics. Boston, MA: Springer; 2015. pp. 37-41

[40] Vavourakis CD, Andrei AS, Mehrshad M, Ghai R, Sorokin DY, Muyzer G. A metagenomics roadmap to the uncultured genome diversity in hypersaline soda lake sediments. Microbiome. 2018;**6**(1):168

[41] Hemmat-Jou MH, Safari-Sinegani AA, Mirzaie-Asl A, Tahmourespour A. Analysis of microbial communities in heavy metals-contaminated soils using the metagenomic approach. Ecotoxicology. 2018;**27**(9):1281-1291

[42] Korzhenkov AA, Toshchakov SV, Bargiela R, Gibbard H, Ferrer M, Teplyuk AV, et al. Archaea dominate the microbial community in an ecosystem with low-to-moderate temperature and extreme acidity. Microbiome. 2019;**7**(1):11

[43] Voorhies AA, Eisenlord SD, Marcus DN, Duhaime MB, Biddanda BA, Cavalcoli JD, et al. Ecological and genetic interactions between cyanobacteria and viruses in a low-oxygen mat community inferred through metagenomics and metatranscriptomics. Environmental Microbiology. 2016;**18**(2):358-371

[44] Riquelme C, Dapkevicius MDLE, Miller AZ, Charlop Z, Brady S, Mason C, et al. Biotechnological potential of Actinobacteria from Canadian and Azorean volcanic caves. Applied Microbiology and Biotechnology. 2017;**101**(2):843-857

Chapter 3

Genotype-specific Patterns of Physiological, Photosynthetic, and Biochemical Responses in Faba Bean Contrasting Pair to Salinity

Imene Rajhi, Bechir Baccouri, Safa Khalifa, Fethi Barhoumi, Moez Amri and Haythem Mhadhbi

Abstract

To understand the salinity tolerance mechanism in faba bean contrasting pair to salinity (cv. Chourouk as sensitive and cv. Najeh as tolerant), we evaluated the effect of high salt concentration (150 mM NaCl) on the photosynthetic, physiological, and biochemical parameters at short and long term of treatment (1 and 6 days, respectively) in the seedling stage. In general, the salinity affects the growth of plants. High salinity decreased all studied parameters, especially transpiration rate (E), stomatol conductance (g_s), net CO_2 assimilation (A), and substomatal CO_2 concentration (C_i), and dramatic changes was registered in cv. Chourouk compared to cv. Najeh. Chlorophyll contents were also affected by salinity, especially in the sensitive variety. In addition, the synthesis of osmolytes (proline) was determinate, to understand whether the osmotic adjustment is a mechanism used by cv. Najeh to tolerate salt stress. Our research suggests that cv. Najeh should be introduced in a crossbreeding program as an elite salt-tolerant germplasm.

Keywords: Faba bean, salinity, photosynthetic parameters, contrasting pair, physiological traits

1. Introduction

One of the most damaging abiotic stimuli that affects crop output and changes in plant growth and development is salt stress. High salinity is thought to affect 20% of the world's arable land and 33% of its irrigated agricultural land, with the total area affected by salinity growing at a pace of 2% per year [1]. Additionally, it has been predicted that by 2050, salinity will affect more than 50% of the entire area currently used for agriculture [2]. In semiarid and arid locations, where other environmental stresses are more prevalent, the effects of salt stress are more severe. For example, insufficient precipitation can create drought, heat can produce excessive surface evaporation, and irrigation with saline water can exacerbate salinity problems [3]. In response to salt stress, various elements of plant development at the germination, vegetative, and reproductive stages are altered because of interactions between physiological, morphological, and biochemical systems [4]. The plant is subject to nutritional deficiency, oxidative stress, ion toxicity, and osmotic stress because of the soil's salinity. The main two reasons of the growth inhibition of salinity were as follows: the high concentration of salt which decrease the ability of plants to adsborb water, and this can lead to the touble in the growth rate (water-deficit effect or osmotic effect); and the ionic effects which due to the diffusion of high amounts of salt in the plant tissues will cause the damage of the cells [5]. However, the level of salt tolerance differs amongt species [6].

The flowering plant Vicia faba, commonly known as the broad bean, fava bean, or faba bean, horse bean, field bean, bell bean, Windsor, or tic bean, is native to North Africa and south-western Asia and is widely cultivated abroad [7, 8]. In warm temperate and subtropical regions, it is grown as a winter annual. While the hardiest European cultivars may withstand winter temperatures as low as −15°C, harder cultivars growing in the Mediterranean region can withstand winter temperatures as low as −10°C [9]. Due to its high protein content, which can range from 20 to 41% depending on variety, faba bean seeds are particularly essential crops [10]. Additionally, they have enough amounts of both carbohydrates and oil, which raises their nutritional value [10]. Due to their great nutritional content and positive health impacts, the consumption of legumes has recently expanded around the world [11–14].

Two types of faba beans, *Vicia faba* L. var. major and *Vicia faba* L. var. minor, are grown in Tunisia, and on average, faba bean cultivation takes up roughly 68 percent of the country's total acreage for growing grain legumes area [15]. The average of dry seed yield of the country is 0.99 t ha^{-1}, which is less than the globally average yield (1.7 t ha^{-1}). However, the production of faba beans in Tunisia varies from year to year because of the lack of cultivars that are resistant to biotic and abiotic challenges (mostly salt). The creation of salt-tolerant cultivars is a crucial method for reducing the detrimental impact of salinity on agricultural productivity. Crop tolerance to salt has long been studied using plant breeding techniques and traditional selection methods. It has been noted that selection based on physiological characteristics can ameliorate salt tolerance in crops better than selection based on agronomic traits [16]. Based on the physiological aspects, Rajhi and his collegues [8] succeed to select a pair of Vicia faba with contrasting behaviour to salinity stress; cv. Chourouk was selected as sensitive cultivar and Najeh was considered as a tolerant one [7, 8]. Thus, the objective of this study is to study the genotype-specific patterns of physiological, photosynthetic, and biochemical responses in faba bean contrasting pair to salinity, grown under high salinity level (150 mM of NaCl).

2. Materiels and methods

2.1 Plant materials

Two Tunisian faba bean cultivars (cv. Najeh and cv. Chourouk) were chosen in this study to understand the physiological and biochemical tolerance to stress salinity. Seeds were offred from the Field Crops Laboratory of the National Institute of Agricultural Research in Tunisia (INRAT).

2.2 Growth conditions

This study was conducted in the greenhouse of the Biotechnology Center of Biotechnology of Borj Cedria under controlled conditions: 23°C, 16/8-h day/night photoperiod, relative humidity of 55–65%, and photosynthetically active radiation of 270 μmol (photon) m^{-2} s^{-1}. Seeds were first surface sterilized by soaking them in a solution of 0.1% mercuric chloride ($HgCl_2$) for 1 minute, followed by a thorough rinsing with sterilized distilled water. They were then planted in humidifying perlite at room temperature (20°C) in the experimental field for 7 days. Then, similar size seedlings were moved into plastic pots filled with a half-strength nutritional solution for 7 days. After that, the seedlings were transferred to a full-strength nutrient solution containing micronutrients (in μM: $CuSO_4$, 1.56; $ZnSO_4$, 1.55; H_3BO_3, 4; $(Na)_2MoO_4$, 0.12; $MnSO_4$, 6.6; $CoSO_4$, 0.12) and macronutrients (in mM: K_2SO_4, 0.7; KNO_3, 24; KH_2PO_4, 0.36; $CaCl_2$, 1.65; $MgSO_4$, 1) [17]. Every week, fresh nutrient solutions were added, and a hydroponic air pump system continuously aerated them at 400 ml/min.

2.3 Salt stress treatment

Plants were subjected to salinity treatment when they reached the four-leaf stage under two different settings: (1) control conditions (0 mM NaCl) and (2) severe salt concentration (150 mM NaCl). By gradually introducing 25 mM of NaCl into the nutritional solution each day, salinity stress was induced. Weekly nutrient solution changes were made to both the treated and control plants. The treated and control plants were both collected individually. Nine replications were considered for each cultivar per treatment.

2.4 Measurements of morphological parameters

The number of leaves, shoot lengths, and root lengths were the morphological parameters that were measured in this study. The distances between the crown and the leaf tip (in cm) and the crown and the root tip (in cm), respectively, were used to calculate the lengths of the shoots and roots. Counting was used to establish how many leaves there were.

2.5 Plant biomass

Treated and control plants were harvested and divided into roots and shoots. On the day of harvest, the weights of the shoots and roots were measured. Each part was dried at 70°C until a constant weight was attained, and then the dry weight was measured.

2.6 Relative water content

According to Barrs and Weatherley's instructions [18], the relative water content (RWC) was calculated. To determine the fresh weight, the topmost fully expanded leaves were weighted (FW). The leaves were then weighted again to obtain the turgid weight after being immersed in distilled water for 24 hours (TW). Following that, the samples were dried at 70°C for 72 hours, and the dry weight was calculated (DW).

The following equation was used to calculate the RWC: RWC (%) = {(FW/DW)/(TW/DW)} *100.

2.7 Gas exchange measurements

The photosynthetic parameters were determined using the youngest fully developed leaf under the following conditions: full sunlight and at 10:00 a.m. under atmospheric CO_2. An open-type and portable photosynthetic system (LC pro+; Bio-Scientific, Great Amwell, Herts, UK) was used to monitor the internal concentration of CO_2 (C_i), stomatal conductance (g_s), net CO_2 assimilation rate (A), and transpiration rate (E).

2.8 Chlorophyll content

Using the Lichtenthaler method [19], the chlorophyll content in different samples was estimated. Young leaves that had just been cut (100 mg) were foated in 5 ml of an acetone solution (80% of acetone) at 4°C and in the dark until the chlorophyll extraction process was complete. The amount of total chlorophyll was determined by measuring absorbance (A) at 645 and 663 nm using a UV-visible spectrophotometer (Jenway 6850 UV–Vis; ColeParmer Ltd., UK). The chlorophyll content was measured as mg g^{-1} as following: total chlorophyll content = 6.45 × A663 + 17.72× A645.

2.9 Electrolyte leakage

According to Dionisio-Sese and Tobita [20], electrolyte leakage (EL) was calculated. Freshly cut leaves were placed in an assay tube containing ultrapure water, and the tubes were then incubated at 32°C for 2 hours. During this time, the solution's

Number	Parameters	Abbreviation
1	Root length	-
2	Shoot length	-
3	Root fresh biomass	-
4	Shoot fresh biomass	-
5	Root dry biomass	-
6	Shoot dry biomass	-
7	Leaf number	-
8	Proline	-
9	Relative water content	RWC
10	Chlorophyll	-
11	Electrolyte leakage	-
12	Transpiration rate	A
13	Stomatal conductance	g_s
14	Net CO_2 assimilation	A
15	Stomatal CO_2 concentration	C_i

Table 1.
Different used parameters.

first electrical conductivity (EC1) was measured using a Metrohm 712 conductometer (Metrohm AG, Herisau, Switzerland). Following the EC1 measurement, the same tubes were incubated in a 90°C oven for 2 hours. Then, the solution was cooled to 25°C, and the second conductivity (EC2) measurement was performed. Electrolyte leakage parameter was calculated using this formula: EL = EC1/EC2 × 100.

In total, 15 parameters were measured (**Table 1**).

In the current investigation, all morphological and physiological data were converted into salt tolerance indices using the method of Zeng et al. [21] and Rajhi et al., [8]. These indices were calculated by dividing the salinity-related observed value by the control averages. The difference between treatment data was estimated using the STATISTICA software and the means of comparison by HSD (higher significant difference) Duncan's test ($p \leq 0.05$).

3. Results

In the current study, 15 physiological, morphological, and biochemical characteristics were used to compare the responses of a contrasting pair of faba bean genotypes to severe salt concentration (150 mM) as well as control conditions.

All the data were converted to relative values, or salt tolerance index. The observation under salinity was divided by the means of the controls to calculate the salt tolerance index. The treated and control plants were grown in identical environmental conditions and harvested simultaneously with and without the addition of NaCl, respectively.

3.1 Effect of salinity on plant length, fresh and dry biomass, and leaf number

Figure 1 describes the effects of salinity stress on the growth characteristics of faba bean plants. In general, this figure shows a significant difference between both cultivars during treatment time. Seedling shoot and root lengths of cv. Chourouk

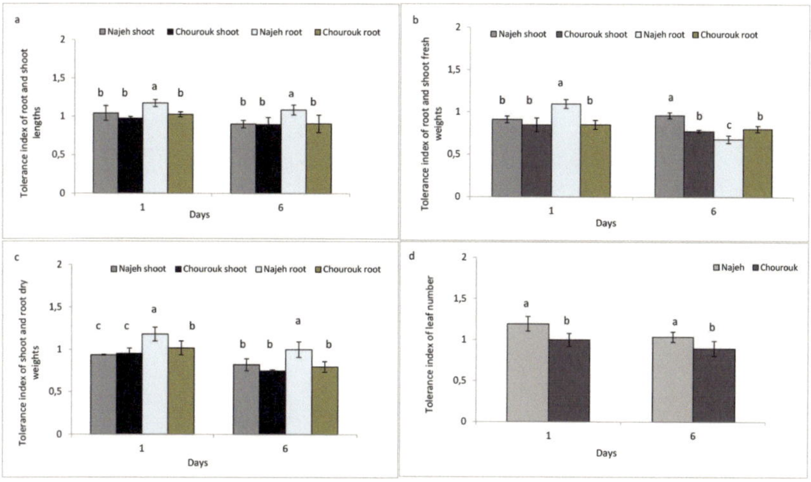

Figure 1.
Index of tolerance of root and shoot lengths (a), index of tolerance of fresh root and shoot weights (b), index of tolerance of dry root and shoot weights (c), and index of tolerance of leaf number (d) of cv. Najeh and cv. Chourouk. All values are means ± SD. The data followed by different letters are significantly different at $p \leq 0.05$.

decreased with the application of stress after 6 days compared to 1 day (**Figure 1a**). However, the length of the root of tolerant cv. Najeh did not affect by salinity applied for short or long time compared to the other cultivar. **Figure 1b** clearely demonstrated that the fresh root biomass of cv. Najeh had the highest value compared to root and shoot of cv. Chourouk at short time of stress application. Nevethless, after 6 days of stress conditions, cv. Najeh exhibited the lowest fresh biomass compared to shoot and root of cv. Chourouk. Furthermore, the dry biomass of cv. Najeh presented the highest value compared to the root of cv. Chourouk after short or long time of stress application (**Figure 1c**). On the other hand, the dry biomass of shoot of cv. Chourouk exhibited the lowest value after 6 days of NaCl supply.

The results of the number of leaves shown in **Figure 1c** indicate that the use of high salinity concentration decreased the number of leaves in stressed plants compared to control ones. Cv. Chourouk had the lowest tolerance index of leaf number compared to cv. Najeh, which presented the highest values after 1 or 6 days of salt treatment.

3.2 Effect of salinity on proline content, total chlorophyll content, electrolytes leackage, and RWC

Proline accumulation is an important mechanism for osmotic regulation under salt stress. In this study, we evaluated proline accumulation profiles in roots and shoots of a contrasting pair of Vicia faba to salinity stress (**Figure 2a**). An increase in proline accumulation was increased in all plant's parts. Shoot and root of cv. Najeh exhibited the highest proline contents either at 1 day or 6 days of salt application. An important accumulation of proline was registered in the root of cv. Najeh after 6 days of tretement compared to cv. Chourouk.

Figure 2b illustrates the impact of salinity on RWC. A significant difference was found between both cultivars in response to 150 mM of salt concentration. The index

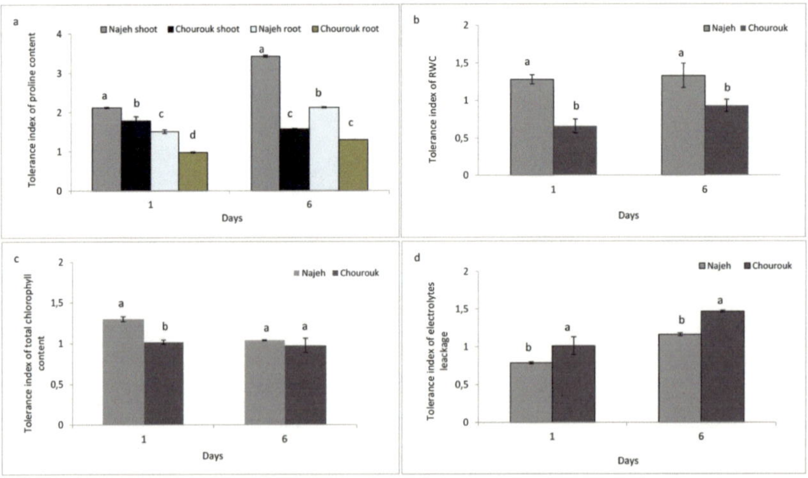

Figure 2.
Tolerance index of proline content (a), RWC (b), total clorophyll content (c), and electrolyte leackage (d) of cv. Najeh and cv. Chourouk. All values are means ± SD. The data followed by different letters are significantly different at $p \leq 0.05$.

of tolerance of RWC of cv. Najeh was not affected by high salinity stress either after 1 or 6 days of treatment. The highest RWC was recorded for cv. Najeh compared with cv. Chourouk.

The chlorophyll content of the leaves of the faba bean cultivars significantly decreased because of exposure to the important amount of NaCl, as evidenced by the results in **Figure 2c**. After 24 hours of salt application, cv. Najeh exhibited the highest value in term of chloroplyll content compared to cv. Chourouk. However, after 6 days of treatmet, both cultivars exhibited the same value of chlorophyll.

In this study, we noted a significant increase in electrolyte leackage in stressed plants (**Figure 2d**). The electolyte leackage increased with the increase of the time of treatment. The best results were observed in cv. Najeh after short and long time of treatment compared to cv. Chourouk, which had the highest value.

3.3 Effect of salinity on photosynthetic parameters

Figure 3 displays the effects of salinity on the leaf gas exchange parameters in faba bean cultivars under treatment condition during different time. In comparison to their respective controls, the index of tolerance of *Ci* of both cultivars did not change after 1 day of salt application, and they exhibited a comparable value (**Figure 3a**). However, after 6 days this parameter was affected by salinity and a large decrease was detected for cv. Chourouk. A significant decrease was also observed for *E* parameter for both cultivars (**Figure 3b**). In addition, this parameter was lower after 6 days compared to 1-day salinity duration. A significatant fluctuation was recorded between both cultivars, and cv. Najeh showed the highest *E* value compared to cv. Chourouk. No signifcant diference was found in g_s values between cultivars (**Figure 3c**). Neverthelees, long salt period affects this parameter. The *A* parameter was also affected by salinity (*A*), especially after 6 days of treatment (**Figure 3d**). However, we did not find a significant difference between both cultivars.

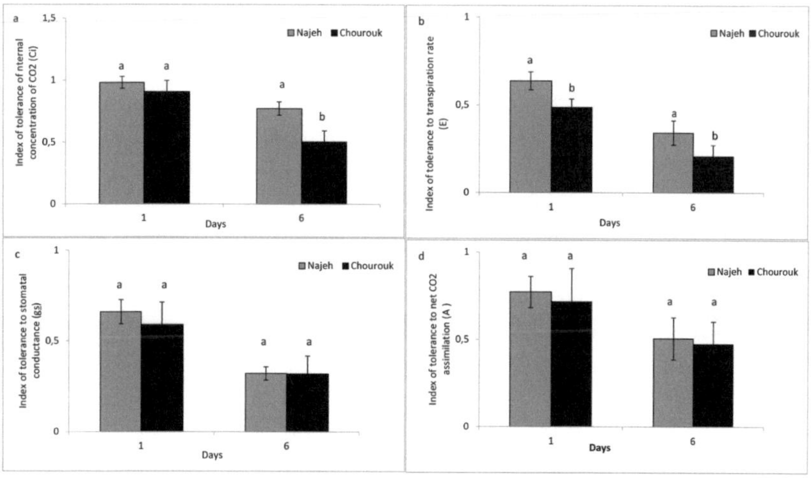

Figure 3.
Tolerance index of internal concentration of CO_2 (a), transpiration rate (b), stomatal conductance (c), and net CO_2 assimilation (d) of cv. Najeh and cv. Chourouk. All values are means ± SD. The data followed by different letters are significantly different at $p \leq 0.05$.

4. Discussion

The number of regions with salt-affected soils is likely to grow in the future years, especially for arid and semiarid regions like Tunisia, which is facing the most severe impacts of salinity stress. A rapidly growing global population will have fewer food options due to limited crop output caused by the degradation of fertile land. The creation of salt-tolerant plants may make it possible to cultivate saline-affected land. The current study was conducted in this context to establish morphological and physiological characteristics that can be used to assess the tolerance of faba bean to salinity and to compare the salt tolerance of two faba bean cultivars with contrasting behaviour to salinity at the seedling stage. Our results showed that salinity negatively affect the lengths, the fresh, and dry biomasses of plants. These results reveal that salinity treatment had an impact on the study plants' growth and biomass development, which are in accordance with the data provided by Tavakkoli et al. [16], Hashem et al. [22], and Dawood and EL-Awadi [23]. This decrease in growth and biomass is caused by the suppression of cell extension and division, production of reactive oxygen species, decrease in mineral intake, hormonal imbalance, and inhibition of enzyme activity [24–26]. NaCl harmful effects on plant metabolism, particularly sensitive plants, cause their slowly growth. It has been discovered that sensitive cultivars are more susceptible to salinity than tolerant ones [24]. Salinity treatment significantly decreased the root and shoot lengths and weights of the cv. Chourouk compared to cv. Najeh. This pattern can be a sign of salt sensitivity. Our findings demonstrate that faba bean plants lost leaves when exposed to salnity stress. Similar results have been reported in diferent plants [27–29]. Our findings demonstrate that cv. Najeh was able to maintain the number of leaves under salinity circumstances compared to control ones, since it exhibited the smallest decrease in leaf number at short and long term of salt treatment.

Different physiological reactions and changes in photosynthetic mechanisms cause the reduction of plant growth that occurs under salt conditions. Stomatol closure caused on a decrease in intracellular CO_2 is thought to be the cause of photosynthetic limitations during salt stress [30]. Our experiment demonstrated that salinity stress significantly affected plant growth by inhibiting A, E, g_s, and C_i parameters. Our results agree with those reported by Mohamed et al. [31] and Alzahrani et al. [32]. Cv. Najeh exhibited less reduction in photosynthetic activity compared to cv. Chourouk, when grown under 150 mM of NaCl. This stability of photosynthetic parameters reflects an ability to maintain photosynthetic capacity and a better tolerance to salinity conditions. Cv. Najeh is considered as a tolerant cultivar, because it exhibited a better photosynthetic performance than the sensitive one.

Relative water content is the most relevant measure of the status of plant water under stress [33]. Cv. Najeh showed a stable and similar value of RWC after 1 or 6 days of treatment, and that may be due to the capacity for water absorption and water status and leaf hydration [34].

The plasma membrane is the main location of ion-specific salt damage [35]. The electrolyte leackage is considered as one of the criteria in the identification of tolerant cultivars to salinity [36]. In this study, the electrolyte leackage was slightely affected in cv. Najeh. However, cv. Chourouk recorded an increased value of this parameter. The same increasing trend was observed in salt-sensitive plants compared to tolerant ones [37].

Proline is the most prevalent endogenous osmolyte that accumulates under different abiotic stresses, such as salinity [38, 39]. It is widely known that specific exogenous proline concentrations control several aspects of plant growth and development under salt stress, such as increases in biomass and productivity [40–42]. When applied as an exogenous compound to crops, proline can improve salt tolerance [43]. Thus, the tolerance of cv. Najeh may be due to the accumulation of proline as osmolyte under salinity stress.

5. Conclusions

In conclusion, the physiological and photosynthetic parameters evaluated in this study demonstrated significant genotypic variation, confirming that variables that really do may be utilized as salinity tolerance screening criteria for faba beans. The examinated cultivars varied significantly in how they responded to salinity stress. However, the physiological characteristics of two cultivars were affected by salt stress to varied degrees, suggesting that these cultivars' resistance to salinity varies. As a result of its capacity to preserve both its photosynthetic system and biomass, our data collectively indicate that cv. Najeh is the best cultivar in surviving salinity-stressed circumstances. The major distinguishing characteristics used to categorize faba bean varieties are thought to be photosynthetic and biomass parameters. The findings of this study demonstrate that several pathways contribute to salinity tolerance. This study specifically demonstrate that cv. Najeh can be classified as a salt-tolerant cultivar, which may be of great interest in future breeding projects for modern cultivar improvement. Finally, our research suggests that cv. Najeh should be introduced in a crossbreeding program as an elite salt-tolerant germplasm.

Author details

Imene Rajhi[1]*, Bechir Baccouri[2], Safa Khalifa[1], Fethi Barhoumi[1], Moez Amri[3,4] and Haythem Mhadhbi[1]

1 Laboratory of Legumes and Sustainable Agrosystems, Biotechnology Center of Borj Cedria, Route Touristique Borj Cedria-Soliman, Hammam-Lif, Tunisia

2 Laboratory of Olive Biotechnology, Biotechnology Center of Borj Cedria, Route Touristique Borj Cedria-Soliman, Hammam-Lif, Tunisia

3 Agro-BioSciences (AgBS), University Mohammed VI Polytechnics (UM6P), Benguerir, Morocco

4 Crop Laboratory, National Institute of Agricultural Research of Tunisia (INRAT), Ariana, Tunisia

*Address all correspondence to: imenrajhi@yahoo.fr

References

[1] Shrivastava P, Kumar R. Soil salinity: A serious environmental issue and plant growth promoting bacteria as one of the tools for its alleviation. Saudi Journal of Biological Science. 2015;**22**(2):123-131

[2] Jamil A, Riaz S, Ashraf M, Foolad MR. Gene expression profling of plants under salt stress. Critical Reviews in Plant Sciences. 2011;**30**(5):435-458

[3] Azevedo Neto AD, Prisco JT, Enéas-Filho J, Abreu CEB, Gomes-Filho E. Efect of salt stress on antioxidative enzymes and lipid peroxidation in leaves and roots of salt-tolerant and salt-sensitive maize genotypes. Environmental and Experimental Botany. 2006;**56**(1):87-94

[4] Akbarimoghaddam H, Galavi M, Ghanbari A, Panjehkeh N. Salinity efects on seed germination and seedling growth of bread wheat cultivars. Trakia Journal of Science. 2011;**9**(1):43-50

[5] Zhang W, Bao G, Tang W, Dai G, Xiao J, Liu J, et al. Physiological response of barley seedlings to salinity and artemisinin combined stresses under freeze-thaw environment. Environmental Science Pollution Research. 2022 [Article ahead of print]

[6] Munns R, Tester M. Mechanisms of salinity tolerance. Annual Review of Plant Biology. 2008;**59**:651-681

[7] Neji I, Rajhi I, Baccouri B, Barhoumi F, Amri M, Mhadhbi H. Leaf photosynthetic and biomass parameters related to the tolerance of Vicia faba L. cultivars to salinity stress. Euro-Mediterranean. Journal for Environmental Integration. 2021;**6**:22

[8] Rajhi I, Moussa S, Neji I, Baccouri B, Chikha M, Mhadhbi H. Photosynthetic and physiological responses of small seeded faba bean genotypes (Vicia faba L.) to salinity stress: Identification of a contrasting pair towards salinity. Photosynthetica. 2020;**58**(1):174-185

[9] Robertson LD, Singh KB, Erskine W, Abd El Moneim AM. Useful genetic diversity in germplasm collections of food and forage legumes from West Asia and North Africa. Genetic Resources and Crop Evolution. 1996;**43**:447-446

[10] Rodino AP, Nistal JH, Hermida M, Santalla M, De Ron AM. Sources of variation for sustainable field pea breeding. Euphytica. 2009;**166**:95-107

[11] Rajhi I, Baccouri B, Rajhi F, Mhadhbi H, Flamini G. Monitoring the volatile compounds status of whole seeds and flours of legume cultivars. Food Bioscience. 2021;**41**:101105

[12] Rajhi I, Boulaaba M, Baccouri B, Rajhi F, Flamini G, Mhadhbi H. Assessment of dehulling effect on volatiles, phenolic compounds and antioxidant activities of faba bean seeds and flours. South African Journal of Botany. 2022a;**147**:741-753

[13] Rajhi I, Baccouri B, Rajhi F, Mhadhbi H, Flamini G. HS-SPME-GC–MS characterization of volatile chemicals released from microwaving and conventional processing methods of fenugreek seeds and flours. Industrial Crops Production. 2022b;**2022**:182

[14] Rajhi I, Baccouri B, Rajhi F, Hammami J, Souibgui M, Amri M, et al. Evaluation of germination effect on volatile compounds of different faba bean cultivars using HS-SPME/GC-MS. Journal of Food Composition and Analysis. 2022c;**112**:104692

[15] Kharrat M, Ouchari H. Faba bean status and prospects in Tunisia. Grain Legumes. 2011;**56**:11-12

[16] Tavakkoli I, Paull J, Rengasamy P, McDonald GK. Comparing genotypic variation in faba bean (Vicia faba L.) in response to salinity in hydroponic and feld experiments. Field Crops Research. 2012;**127**:99-108

[17] Vadez V, Rodier F, Payre H, Drevon JJ. Nodule permeability to O2 and nitrogenase-linked respiration in bean genotypes varying in the tolerance of N2 fxation to P defciency. Plant Physiology and Biochemistry. 1996;**34**:871-878

[18] Barrs C, Weatherley PE. A re-examination of the relative turgidity technique for estimating water defcit in leaves. Australian Journal of Biological Sciences. 1968;**15**:413-428

[19] Lichtenthaler HK. Chlorophylls and carotenoids: Pigments of photosynthetic biomembranes. Methods in Enzymology. 1987;**148**:350-382

[20] Dionisio-Sese ML, Tobita S. Antioxidant responses of rice seedlings to salinity stress. Plant Science. 1998;**135**:1-9

[21] Zeng L, Shannon MC, Grieve CM. Evaluation of salt tolerance in rice genotypes by multiple agronomic parameters. Euphytica. 2002;**127**:235-245

[22] Hashem A, Abd Allah E, Alqarawi A, El-Didamony G, Alwhibi M, Egamberdieva D, et al. Alleviation of adverse impact of salinity on faba bean (Vicia faba L) by arbuscular mycorrhizal fungi. Pakistan Journal of Botany. 2014;**46**(6):2003-2013

[23] Dawood MG, El-Awadi ME. Alleviation of salinity stress on Vicia faba L plants via seed priming with melatonin. Acta Biol Colomb. 2015;**20**(2):223-235

[24] Ahmad P, Ahanger MA, Alyemeni MN, Wijaya L, Egamberdieva D, Bhardwaj R, et al. Zinc application mitigates the adverse efects of NaCl stress on mustard [Brassica juncea (L.) czern and coss] through modulating compatible organic solutes, antioxidant enzymes, and flavonoid content. Journal of Plant Interactions. 2017;**12**(1):429-437

[25] Ashraf MA, Ashraf M, Ali Q. Response of two genetically diverse wheat cultivars to salt stress at diferent growth stages: Leaf lipid peroxidation and phenolic contents. Pakistan Journal of Botany. 2010;**42**(1):559-565

[26] Munns R. Physiological processes limiting plant growth in saline soils: Some dogmas and hypotheses. Plant, Cell & Environment. 1993;**16**(1):15-24

[27] Abdul Qados AMS. Efect of salt stress on plant growth and metabolism of bean plant Vicia faba (L.). Journal of the Saudi Society of Agricultural Sciences. 2011;**10**:7-15

[28] Karen W, Anthony RY, Timothy JF. Efects of salinity and ozone, individually and in combination on growth and ion contents of two chickpea (Cicer aritinum L) varieties. Environmental Pollution. 2002;**120**(2):397-403

[29] Raul L, Andres O, Armado L, Bernardo M, Enrique T. Response to salinity of three grain legumes for potential cultivation in arid areas (plant nutrition). Soil Science & Plant Nutrition. 2003;**49**(3):329-336

[30] Misra AN, Sahu SM, Misra M, Singh P, Meera I, Das N, et al. Sodium chloride induced changes in leaf growth, and pigment and protein contents in

two rice cultivars. Biologia Plantarum. 1997;**39**:257-262

[31] Mohamed AA, Eichler-Löbermann B, Schnug E. Response of crops to salinity under Egyptian conditions: A review. Landbauforsch Volk. 2007;**2**(57):119-125

[32] Alzahrani SM, Alaraidh IA, Migdadi H, Alghamdi S. Physiological, biochemical, and antioxidant properties of two genotypes of Vicia faba grown under salinity stress. Pakistan Journal of Botany. 2019;**51**:3

[33] Yamasaki S, Dillenburg LR. Measurement of leaf relative water content in Araucaria angustifolia. Revista Brasileira de Fisiologia Vegetal. 1999;**11**:69-75

[34] Siddiqui MH, Al-Khaishany MY, Al-Qutami MA, Al-Whaibi MH, Grover A, Ali HM, et al. Response of diferent genotypes of faba bean plant to drought stress. International Journal of Molecular Sciences. 2015;**16**:10214-10227

[35] Mansour MMF, Salama KHA. Cellular basis of salinity tolerance in plants. Environmental and Experimental Botany. 2004;**52**:113-112

[36] Ashraf M, Ali Q. Relative membrane permeability and activities of some antioxidant enzymes as the key determinants of salt tolerance in canola (Brassica napus L.). Environmental and Experimental Botany. 2008;**63**:266-273

[37] Lutts S, Kinet JM, Bouharmont J. NaCl-induced senescence in leaves of rice (Oryza sativa L.) cultivars difering in salinity resistance. Annals of Botany. 1996;**78**:389-398

[38] Slama S, Bouchereau A, Flowers T, Abdelly C, Savouré A. Diversity, distribution and roles of osmoprotective compounds accumulated in halophytes under abiotic stress. Annals of Botany. 2015;**115**:433-447

[39] Szabados L, Savouré A. Proline : A multifunctional amino acid. Trends in Plant Science. 2010;**15**:89-97. DOI: 10.1016/J.TPLANTS.2009.11.009

[40] Huang Y, Bie Z, Liu Z, Zhen A, Wang W. Protective role of proline against salt stress is partially related to the improvement of water status and peroxidase enzyme activity in cucumber. Soil Science & Plant Nutrition. 2009;**55**:698-704

[41] Nawaz K, Talat A, Hussain K II, Majeed A. Induction of salt tolerance in two cultivars of Sorghum (*Sorghum bicolor* L.) by exogenous application of proline at seedling stage. World Applied Science Journal. 2010;**10**:93-99

[42] Wu G-Q, Feng R-J, Li S-L, Du Y-Y. Exogenous application of proline alleviates salt-induced toxicity in sainfoin seedlings. Journal of Animal Plant Science. 2017;**27**:246-251

[43] Heuer B. Role of proline in plant response to drought and salinity. In: Pessarakli A, editor. Handbook of Plant and Crop Stress. 3rd ed. Boca Raton: CRC Press; 2010. pp. 213-238

Chapter 4

Antioxidant Effective Aromatic Compounds

Hulya Celik and Kader İlhan

Abstract

Systems that destroy the effects of free radicals are called antioxidants. Today, reliability concerns on synthetic antioxidants are increasing. Therefore, the interest of the health and food industry in aromatic plants and natural antioxidants obtained from these plants has also increased. Aromatic plants have been used in many fields, such as food, medicine, cosmetics, and spices since the beginning of human history. Interest in plants, which are natural sources of antioxidants, has increased in recent years. Due to this interest, studies on the use of aromatic plants as natural antioxidant sources continue to increase day by day. As a result of research on this subject, it has been shown that aromatic plants contain many phytochemical compounds with high antioxidant activity. Among these compounds, especially phenols stand out as a secondary metabolite group and play an important role in the total antioxidant activity of medicinal plants. In fact, studies have shown that there is a positive relationship between the total phenol content of medicinal plants and aromatic plants and their antioxidant capacity. Phenolic compounds are obtained from plants and contain one or more hydroxyl groups in their aromatic rings. These biologically active components in herbal essential oils have been used as therapeutic agents, as they are natural sources of antioxidants, inactivate free radicals, reduce oxidative stress, and have been implicated in the pharmaceutical, cosmetic, and food research fields. Thymoquinone is the main active ingredient of Nigella sativa L. It has been used in the treatment of various animal and human diseases for hundreds of years in world history. Thymol and carvacrol are the main components of known thyme, and its derivatives and are widely used in pharmacology. In this review, the antioxidant activities of some aromatic plants, whose importance is increasing day by day, and thymoquinone, thymol, and carvacrol phenolic compounds that can be used instead of synthetic antioxidants in the food industry are mentioned.

Keywords: aromatic compounds, antioxidant, biologically active, pharmacology

1. Introduction

The development of science and technology, environmental pollution, and many other factors have exposed us to various toxic substances. The number of human diseases caused by these toxic substances is increasing day by day. These diseases can be solved by controlling free radicals. Free radicals are the main factor provoking the occurrence of these diseases.

Many systemic diseases can occur in our fully functioning bodies throughout our lives due to the proliferation of free radicals after a certain age. As we get older, our defense mechanism weakens and the body's free radical balance is disturbed. Therefore, it is important to consume natural foods that contain antioxidants to restore balance. Some compounds in plants have the effect of delaying the onset of this process. These compounds, which have been studied for a long time, are called antioxidants [1].

Antioxidants are compounds that protect cells from the harmful effects of free radicals. Recently, in the field of medicine, new methods to treat diseases are being investigated, while efforts are made to maintain a healthy life and prevent diseases. One of the most important issues, in this case, is natural antioxidants [2].

The number of studies investigating the use of herbs and spices as natural sources of antioxidants is increasing. The pharmacological properties of vegetable essential oils and their components have been examined and their use in the medical, cosmetic, and industrial fields has been shown to be beneficial. The use of natural antioxidant substances obtained from vegetable materials in place of synthetic antioxidants and preservatives has created an intense interest in natural antioxidants.

First, spices are used in the treatment of diseases, in other words, aromatic plants are as old as human history. After a while, it began to be used in religious ceremonies and in the production of fragrance agents. Protecting food; the field of use has improved with flavoring and other uses [3].

The protective effects of many spices with antioxidant effects in foodstuffs were looked at and compared with other antioxidant properties. The antioxidant effect is determined by the values of peroxide and is especially measured at certain time intervals. Members of the Labiatae family, such as rosemary, sage, thyme, and mint, have been compared with spices with other synthetic antioxidants and their antioxidant effects have been found to be higher.

The most important groups of natural antioxidants are phenolic substances. These substances, which are seen in all parts of plants, are polyphenolic components, the most common among vegetable phenolic antioxidants: Favonoites, cinnamic acid derivatives, coxins, tocopherols, and phenolic acids. They are known to protect substances found in nutrients that can be easily oxidized from oxidation. Aromatic plants, which have long been used as additives to increase the properties of nutrients, such as smell and taste, are therefore becoming increasingly important. In this review, information will be given about natural antioxidants and some aromatic plants that are considered one of the largest sources of antioxidants.

Although oxygen is indispensable for human life, some types of reactive oxygen produced during normal metabolism have the potential to cause intense damage to the body. Atoms or molecules that hold an uncommon electron are called free radicals. In response to the harms of reactive oxygen species, different natural defense systems in the body control free radicals. Antioxidants prevent oxidation caused by free radicals and are also capable of capturing and stabilizing free radicals. Antioxidants delay aging, they achieve this by binding free radicals to themselves or neutralizing them, minimizing possible damage to the body. The most commonly used antioxidants on your day include beta-carotene, vitamins C, E, lycopene, coenzyme Q-10, selenium, zinc, and manganese, as well as organic and inorganic substances [4].

2. Classification of antioxidants

Antioxidants help the body's defense mechanism against the negative effects of free radicals, so today it has gained great importance in terms of health.

In this respect, when we consider the classification of antioxidants both in terms of health and food, we can collect them under two headings. **Table 1** given in the first is natural antioxidants and the second is synthetic antioxidants [5].

Antioxidants			
Natural Antioxidants			Sentetic Antioxidants
Enzymatic	Non-Enzymatic		BHT, BHA
SOD	Endojen	Eksojen	Troloks
Gutatyon peroksidaz	Glutatyon	Vitamin E	Askorbil palmiat
Glutathione reductase	Serüloplazmin	Beta-karoten	Propil Gallat
Glutatyon-S-transferaz	Bilirubin	Ascorbic acid	Tersiyel Bütil
	Albumin	Flavanoidler	Hidrokinon

Table 1.
Classification of antioxidants.

2.1 Mechanism of action of antioxidants

According to the chain reaction theory, the substance (lipid molecule) activated by energy absorption is oxidized by combining with oxygen, and the activated peroxide molecules formed in this way continue autooxidation by transferring their energy to other oxidizable molecules of the substance. With the use of antioxidants, activation energy is used by the antioxidant molecule, it is not able to transfer this energy to other molecules. With the intervention of the antioxidant molecule, many molecules of the substance, which can be autoxidized, are no longer oxidized, that is, oxidation is slowed down and partially stopped.

$$R\bullet + AH \rightarrow RH + A\bullet$$

$$RO\bullet + AH \rightarrow ROH + A\bullet$$

$$OH\bullet + AH \rightarrow H2O + A\bullet$$

$$ROO\bullet + AH \rightarrow ROOH + A\bullet$$

$$A\bullet + \text{The} \rightarrow AO$$

The active molecule of the antioxidant does not transfer its energy to fat molecules, it is usually oxidized to inactive molecules (AH: Antioxidant molecule, A•: Active antioxidant molecule, AO: Inactive antioxidant molecule).

Antioxidants to be used in foods should have some properties. Some of them include:

- It must be harmless to human health,
- It should be used in very small quantities so that it does not increase the cost,

- The natural smell, appearance, and taste of food should not be disturbed,
- It must dissolve in the substance, it will protect or mix thoroughly,
- It should not lose its effect during normal production [6].

2.2 Aromatic compounds

Plants that have smell and taste properties and are also used as medicines due to their therapeutic properties are called aromatic plants. Plants and essential oils have been used to obtain aromatic foods and beverages since the beginning of human history. Their use in hiding smells, attracting the attention of others, controlling health problems, and providing welfare to people and animals shows a cultural and economic status. Essential oils are usually liquid, transparent, multicolored, and complex. The existing compounds they contain are volatile. These compounds are characterized by a strong odor and are secondarily synthesized to protect plants from microbes and insects [7].

2.3 Antioxidant activities of aromatic plants

The antioxidant activity of aromatic plants is related to the phenolic compounds in their structure. The most abundant of these compounds are flavonoids, phenolic acids, and phenolic terpenes. The antioxidant effect of phenolic compounds is due to such reasons as clearing free radicals, forming compounds with metal ions (metal chelation), and preventing or reducing singlet oxygen formation.

The compounds can provide the hydrogen contained in hydroxyl groups in their aromatic rings to prevent lipids and other biological molecules from being oxidized by free radicals. Flavonoids and other phenolic compounds are found mainly in the leaves, flowers, and woody parts of plants. For this reason, aromatic plants are often used as medicines by drying the leaf and flower parts or as essential oil extracts obtained by methods, such as extraction and distillation. Since the chemical compositions of aromatic plants vary depending on many factors, their antioxidant effects will also vary. Akgül and Ayar examined the antioxidant effects of 31 aromatic plants grown in sunflower oil in Turkey and determined that rosemary followed by rosemary and bone to the strongest antioxidant [8].

2.3.1 Rosemary

Rosemary is a valuable essential oil and spice plant from the Lamiaceae family. Important chemical components in rosemary include carnisol, rosmanol, geraniol, pinene, limonene, apigenin, naringenin, luteolin, rosmarinic, vanilic, caffeic acid. Thanks to the polyphenolic components found in rosemary and cyclic diterpene diphenols, carnosolic acid, carnosol, epirustol, rosmanol, iszorosmanol, rosmarinic acid, and hisperidine, antioxidant activity is high. It is used in foods as an antioxidant or natural preservative. Carnosic acid can be used in the treatment of Alzheimer's disease because it protects the brain against free radicals. Rosemary, which is also consumed as a spice, is also known to have a protective effect against diseases. High antioxidant capacity from rosmarinik acid and carnosol is obtained in many ways by civilized dust extract and these powder extracts are used in foods. Essential oils obtained from this plant are used to prevent oxidation caused by fat and protein degradation in meat and meat products. It is stated that carnasol and carnocyclic acid in particular have higher antioxidant activity than BHA, BHT, and propyl galat, which are used as synthetic antioxidants in membrane lipid peroxidation [9].

Polyphenolic compounds	Chemical structure
Carnosic acid	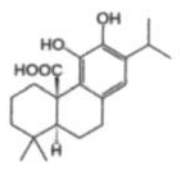
Karnosol	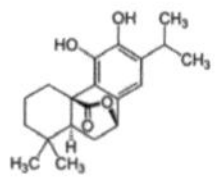
Rosmanol	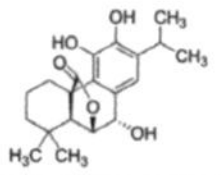

2.3.2 Salvia officinalis *L.*

Sage is another aromatic plant that has a significant antioxidant effect. The most important phenolic compounds in its structure, such as rosemary, are carnosol, carnosic acid, rosemary aldehyde, rosmarinol, epirosmarinol, and methyl carnosine [11].

Sage contains many polyphenol compounds, including phenolic acids and flavonoids. Some of these are caffeic acid and its derivatives, rosmarinic acid, salvianolic acid, luteolin, apigenin, kaempferol, and quercetin. At the same time, sage is a rich essential oil plant, mostly terpenoids. Its antioxidant activity is very high due to the α-thujon, 1,8-cincole, and camphor components in sage essential oil. Medicinal sage is more important than other types of sage due to its high antioxidant activity.

Its high antioxidant activity is mainly due to its structure, which includes sage kumarin, rosmarinic acid, salvianolic acid, and carnocyclic acid. In another study, the components found to have a high antioxidant effect were carnosol, rosmarinic acid, carnocyclic acid, caféic acid, and rosmarinol. These components are reportedly effective in free radical sweeping action and superoxide dysmutase sweeping action. Research has shown that it is also good for forgetfulness and Alzheimer's disease, especially due to its high antioxidant activity. Studies have shown that it has a high-cleansing effect on free radicals and a high effect on DDPH [10, 11].

Phenolic compounds	Chemical structure
Carnosic acid	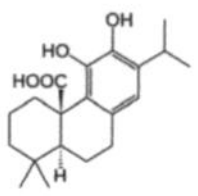
Karnosol	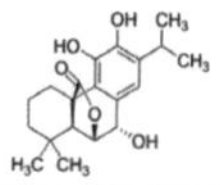
Rosmanol	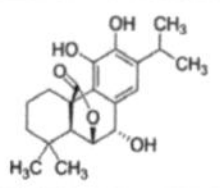

2.3.3 Melissa officinalis *L.*

Son grass (M. officinalis) is a plant belonging to the genus Honeybabagiller, which grows naturally in Turkey. It is a valuable essential oil plant and its leaves are used. The most important components of Calendula essential oil are nerolal (cytral), cytronelal, cytronelol, nerol acetate, isogeraniol, and geranyl acetate. Sage also contains phenolic substances, such as rosmarinic acid, cafleic acid, gallic acid, ramnosid, luteolin, which provide antioxidant properties. Since ancient times, it has been widely used as a sedative, diuretic, diastolic, and analgesic in diseases of the digestive system of people. The antioxidant activity of elderberry flower is related to the cytonellal and nervous components and phenolic components in the essential oil. Some research have used it as an antioxidant, especially as a preservative for high-fat foods. Also due to its high antioxidant capacity, son grass is added to foods with a high-fat content as a protective agent against the oxidation of polyunsaturated fats [12].

Phenolic compounds	**Chemical structure**
Rosmarinic acid	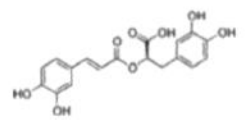
Caffeic acid	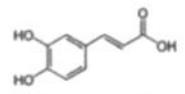
Luteolin	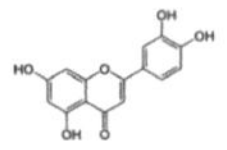

2.3.4 Thyme

Thyme is a valuable essential oil and spice plant from the Lamiaceae family. Especially in essential oils, essential oils containing carvakrol/thymol components are more valuable. The main component of thyme type thyme essential oil is thymol. Especially in Turkey, plants belonging to the genus Origanum are collected and oregano belonging to the genus Origanum is called coral reef. The main component of thyme essential oil is carvakrol. There are also flavonoids hydroxycinnamic acid, hydroxybenzoic acid, rosmarinic acid, apigenin, and luteolin. The antioxidant activity of oregano is mainly due to its essential oils. It is mainly used as an antioxidant to prevent food from spoiling. In terms of human health, herbal tea or essential oil is used for upper throat infections, stomach problems, and antibacterial and antifungal purposes [13].

Essential oil	**Chemical structure**
Karvakrol	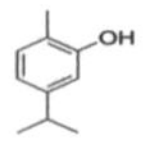
Timol	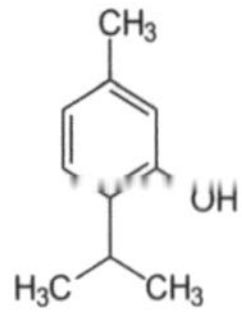

*2.3.5 Green tea (*Camellia sinensis*)*

Green tea is an uncontested type of tea. It is a good source of antioxidants due to the vitamin E in green tea. The highest polyphenols in green tea are catechins and theaflavins. Green tea is rich in flavonoids, including catechins and catechins derivatives. The main catechins in green tea are epigallocatechin gallate (EGCG), epigallocatechin (EGC) and epigallocatechin (EGC), catechins (EC), and epicatechin gallate (ECG). The highest antioxidant effect among catechins. The main flavonols found in green tea are quercetin, kaempferol, myricetin, and rutin. Green tea polyphenols are powerful antioxidants that combine active oxygen and nitrogen substances and indirectly show antioxidant activity by triggering the synthesis of intracellar (endogenous) antioxidant enzymes, such as superoxide dysmutase and glutathione dystase, glutathione-s-reductase, catalase, and kinon reductase. Thanks to these effects, green tea can prevent lipid peroxidation and damage to DNA structure. According to reports, the epigallocatechin gallate in green tea may have a protective effect on neuronal diseases, such as Alzheimer's and Parkinson's by regulating free radical clearing, iron-binding activity, and antioxidant enzymes [14].

Polyphenols	**Chemical structure**
Kateşin	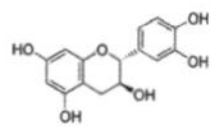
Theaflavin	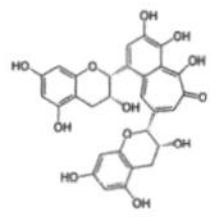

*2.3.6 Likapa (*Vaccinium sp.*)*

Likapa, also known as blueberries, belongs to the family Fundagiller and belongs to the genus blueberries. It is found naturally in my country but is not planted. In our country, blueberries, ligarba, blueberries, morsivit, bush strawberries, and trabzonspor tea are called blueberries in foreign countries. There are four different varieties of natural blueberries in our country, and breeding research is underway well-known species are *Vaccinium myrtillus L.* and *Vaccinium arctostaphylos L.* The parts used are leaves, flowers, and fruits. Phenolic compounds, such as lycapa, chlorogenic acid, kersetin, kaempferol, myrisetin, proantocyanidines, catechins, epicatechin, resveratrol, and anthocyanins have a strong antioxidant capacity as they are a good source of organic acid, tannins, and vitamin C. The fruit does not contain sodium but contains a high percentage of potassium, phosphorus, and calcium. It is a fruit with high nutritional value and strong antioxidant potential, as it contains a high content of anthocyanins, flavonoids, and polyphenol compounds. It has more antioxidant effects than other fruits. In recent years, it has been reported that the consumption of likapa has gradually increased to reduce pathologies associated with oxidative stress [15].

Phenolic compounds	Chemical structure
Kuersetin	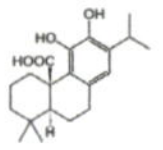
Mirisetin	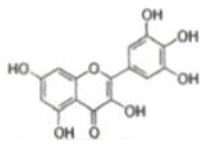
Chlorogenic acid	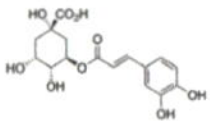

3. Conclusion

In this study, it was determined that substances and plants containing aromatic structures exhibit high antioxidant activity. Research has shown that there is a positive relationship between the phenol content of medicinal and aromatic plants and their antioxidant capacity. Phenolic compounds are obtained from plants and contain aromatic rings and hydroxyl groups. In this review, information was given about the antioxidant activities and general properties of some aromatic plants, which are becoming increasingly important.

Acknowledgements

This study was prepared from Kader İlhan's Aromatic Compounds class research assignment.

Author details

Hulya Celik* and Kader İlhan
Faculty of Pharmacy, Basic Pharmaceutıcal Scıences Department/Fundamental Scıences of Pharmacy, Ağrı Ibrahim Cecen University, Agri, Turkey

*Address all correspondence to: hycelik@agri.edu.tr

References

[1] Payan A. Antioxidant Extract from Grape Fruit and Seeds [thesis]. Konya: Selcuk University, Institute of Natural and Applied Sciences; 2007

[2] Buzzes H. Antioxidant Activity and Total Phenolic Substance Determination of Areceae Arum Dioscarides Seeds Grown in Antalya [thesis]. Konya: Selçuk University, Institute of Science; 2009

[3] Akgül A. Baharat bilimi ve teknolojisi. Ankara: Gıda teknolojisi derneği yayınları. Yayın no: 15; 1993. p. 451s

[4] Ji LL. Antioxidants and oxidative stress in exercise. Proceedings of the Society for Experimental Biology and Medicine. 1999;**222**(3):283-292

[5] Yavaşer R. Doğal ve Sentetik Antioksidan Bileşiklerin Antioksidan Kapasitelerinin Karşılaştırılması [thesis]. Aydın: Adnan Menderes Üniversitesi, Fen Bilimleri Enstitüsü; 2011

[6] Maddox DN. Role of the gallates as foodstuff antioxidants. International Flavours and Food Additives. 1976;7: 117-119

[7] Anonim, Medicinal and Aromatic Plants Working Group-ECP/GR. 2005

[8] Akgül A, Ayar A. Yerli baharatların antioksidan etkileri. Doğa - Turkish Journal of Agriculture and Forestry. 1993;**17**:1061-1068

[9] Nieto G, Ros G, Castilla J. Antioxidant and antimicrobial properties of rosemary (Rosmarinus officinalis, L.): A review. Medicine. 2018;**5**(98):2-13

[10] Lopresti LA. Salvia (sage): A review of its potential cognitive-enhancing and protective effects. Drugs in R&D. 2017; **17**:53-64

[11] Cuvelier ME, Berset C, Richard H. Antioxidant constituents in sage (Salvia officinalis). Journal of Agricultural and Food Chemistry. 1994;**42**:665-669

[12] Miraj S, Kopaei R, Kiani S. Melissa officinalis L: A review study with an antioxidant prospective. Journal of Evidence-Based Complementary & Alternative Medicine. 2017;**22**(3): 385-394

[13] Gutierrez GEP, Salas MAP, Lopez LN, Mendoza MSC, Olivo GV, Heredia JB. Flavonoids and phenolic acids from oregano: Occurrence. Biological Activity and Health Benefits, Plants. 2017;**7**(2):2-23

[14] Weinreb O, Mandel S, Amit T, Youdim MBH. Neurological mechanisms of green tea polyphenols in Alzheimer's and Parkinson's diseases. Journal of Nutritional Biochemistry. 2004;**15**(9): 506-516

[15] Seyhan S. Determination of Ellagic Acid and Resveratrol Amounts by HPLC Method in Blueberry Species Grown in Turkey [thesis]. Istanbul: Marmara University, Institute of Science; 2013

Chapter 5

Extremophiles and Limits of Life in a Cosmic Perspective

Nawab Ali, Muhammad Nughman and Syed Majid Shah

Abstract

Extremophiles are one of the most extreme entity on planet earth which can withstand many harsh conditions considered lethal for other life form of terrestrial life. From an evolutionary prospective, extremophiles are considered to be primitive cells that used to live in the early earths harsh environment living on this planet since billions of years, it can be found in almost in any environmental conditions on our planet. There are many established valuable uses of these extremophiles and particularly their bioactive compounds. The enzymes produced by extremophiles have significant applications in different industries like detergent, food, feed, starch, textile, leather, pulp and paper, and pharmaceuticals This chapter discuss extremophile, their survival mechanism and astrobiology, discussing life in a cosmic prospective.

Keywords: extremophiles, bioactive compounds, limits of life, astrobiology

1. Introduction

Over the period of the last century, the conditions at which life is able to flourish have been pushed in every possible direction, expanding to include wider range of temperature, pH, pressure, radiation, salinity, energy, and nutrient limitation. This has allowed for the discovery of life in previously unimaginable environments. Microorganisms have the ability to thrive in different harsh conditions on earth and in space, which include microgravity, high radiation, vacuum pressure, and extremely diverse temperature [1, 2]. Extremophiles are those organisms which can withstand many harsh conditions which are considered lethal for other life form of terrestrial life. They are able to flourish in extremely higher and lower temperature, as well as acidic and alkaline environments. Many extremophiles can also thrive in different organic solvents, heavy metals and hazardous waste [3]. Space has been often called "The Final Frontier," where most of the universe's conditions are hazardous for human habitation [4]. Extremophiles are essential for discovering biosignatures that can be used to find habitable environments beyond the earth. Different ecological habitat of earth has similarity to that of other planets in term of nutrient composition and biogeochemistry [5]. The aim of this chapter is to discuss extremophiles, importance of their bioactive compounds. Later on, the chapter is focused on life in extreme conditions, limits of life and astrobiology, discussing life in a cosmic prospective.

2. Extremophilic microorganisms

The term extremophile was first coined by McElroy in the year 1974 [6]. Most extremophiles are classified within the archaea, bacterial, and eukaryotic kingdoms [7]. Around a decade ago extremophiles were considered exotic organisms that were only investigated by few researchers around the world. Now, it has become a promising field for enzymologists to explore and utilize these microorganisms in a variety of industrial applications [8]. Studies on extremophiles have advanced substantially over the past twenty years where first international congress on extremophiles was held in Portugal (1996) and the peer-reviewed scientific journal "Extremophiles" was launched in 1997. In addition, in 2002, the "Worldwide Society for Extremophiles" (ISE) was established as an international organization for the purpose of facilitating the sharing of knowledge and expertise in the rapidly expanding field of study on extremophiles [9].

Extremophilic organisms have the ability to grow in such environments which are considered inhospitable to other life forms. These conditions include extreme hot and cold environments, and highly alkaline and acidic environments. Several extremophiles can also thrive in different hazardous waste, organic solvents and heavy metals. It's been discovered that extremophiles can also live in more than ten kilometers deep in the ocean and 6.7 kilometers deep in the earth's crust; in conditions ranging from 0 to 12.8 pH, in temperatures from 122°C to 20°C; and at pressures of up to 110 megapascals (MPa) [10]. Some organisms are not only stable at harsh conditions, but they need them to survive. Extremophiles are classified into different classes based on the conditions in which they thrive, they are thermophiles, which can withstand extreme temperature, psychrophiles, which can withstand extreme low temperature, acidophiles and alkaliphiles, which can withstand extreme higher and lower pH, barophiles, which can withstand higher pressure and halophiles which can withstand higher salt concentration [11]. Extremophiles are sometimes considered as polyextremophiles which mean that they can withstand multiple harsh conditions. Some hot springs have diverse harsh conditions i.e., they are alkaline and basic at the same time and also have higher amount of heavy metals. Different hypersaline lakes are extremely alkaline, and the deep oceans have also diverse harsh condition like cold, oligotrophic and high pressure [12].

Extremophiles have attracted a lot of attention because they are able to catalyze reactions despite harsh conditions and also have significant applications in industries [13–15]. Now researchers are focusing more on genetic engineering of enzymes to improve their activity and to make them interesting candidate in industrial and biotechnological processes. This is because extremozymes are more difficult to isolate than other types of enzymes. Microbes with desirable industrial traits are genetically engineered by a large number of research organizations and companies across the world, and these microorganisms has been used in different industrial processes [16]. Industrial enzymes had a market value of more than one billion dollars in 2010, and it was projected that this value will increase to five billion dollars in 2021 at 4% growth per year [17]. This was achieved earlier, as they hit $5.5 billion in 2018, and various research associations now estimate that they will reach $7.0 billion by 2023 [18, 19]. Extremozymes are the most promising options to take into consideration to meet the ever-increasing demand on the global market [14]. Extremophiles can produce different industrially important enzymes which are stable at different harsh conditions. We have reviewed different extremophiles, there habitat, growth characteristics, bioactive compounds and their industrial importance as shown in the **Table 1**.

Extremophile	Habitat	Growth characteristics	Bioactive compounds	Species	Application	Reference
Thermophiles	Hydrothermal vents, Hot springs	Growth temperature of more than 80°C	Amylase	*Bacillus mojavensis*	Glucose fructose for sweetness	[20]
			DNA polymerase	*Thermus aquaticus (Taq)*	PCR, Diagnostics, molecular biology	[21]
			Xylanases	*Bacillus tequilensis*	Biorefinery, food	[22]
Halophiles	Salt mines, marine, soil	High salt Concentration i.e., more than 5 M	α- amylase	*Halothermorthrix orenii*	Bio catalysis	[23]
			Protease	*Pseudoalteromonas Sp.*	Detergent, Peptide synthesis	[24]
Psychrophiles	Antarctica soil, deep ocean, Mariana trench	Low temperature i.e., below 15°C	β- Galactosidase	*Arthrobacter species C2*	Biorefinery, ethanol production	[25]
			Lipase	*Pyschrobacter okhotskensis*	Food, cosmetics	[26]
Alkaliphiles	Salt mines, soil, soda lakes	Higher pH i.e., above 9	Protease	*Bacillus firmus*	Detergent, food and feed	[27]
			Cellulase	*Bacillus subtilis*	Fermentation of wine	[28]
Acidophiles	Man-made riches, hot springs	Lower pH i.e., lower than 3	Endocellulase	*Thermomonospora (Actinomycetes)*	Leather industries	[29]
Piezophiles	Deep seam, Mariana trench	Pressure above 110 MPa	Lipase	*Colwellia hadaliensis BNL-1*	Food processing	[30]
			Chymotrypsin	*Shewanella benthica strains*	Pharmaceutical industry	[31]
Radiophiles	High UVR altitudes, Mountains	UVR above 110 nm	Mycosporine-like amino acids (MAAs)	*Deinococcus radiodurans*	Diagnostics, bioremediation and therapeutics	[32]

Table 1.
Overview of different extremophiles, their habitat and application of their bioactive compounds.

2.1 Thermophilic microorganisms and their importance

Researcher have shown a lot of interest in extremophiles specifically thermophilic bacteria, due to its ability of growing at above 50°C and producing thermostable enzymes [33]. Thermophiles are categorized into 3 classes: moderate thermophiles which have the ability to withstand 50–60°C, extreme thermophiles which can withstand 60–80°C and hyperthermophiles which have the ability to withstand 80–110°C. Thermophiles have been found in different environments, from hot springs to the deep ocean. Members of *Ascomycete*, *Zygomycete* families of fungi and archaean genera *Pyrobaculum, Pyrodictium, Pyrococcus*, and *Melanopyrus* are the organisms that are capable of growing at temperatures ranging from 103 to 110°C [34], whereas *Thermotoga maritime* and *Aquifex pyrophilus* bacterial species have the ability to grow at extremely high temperature ranging from 90 to 95°C [35, 36]. Utilizing enzymes that are both stable and active at higher temperatures is important for a number of purposes, the primary one is that these enzymes are better suited for conducting biotechnological activities at higher temperatures [8].

Many polymers-degrading enzymes (such as amylases, cellulases, chitinases, pectinases, pullulanases, and xylanases), as well as proteases, isomerases, esterases, lysases, phytases, dehydrogenases, and DNA-modifying enzymes, have been characterized from extremely thermophilic and hyperthermophilic microorganisms [37]. Because of their ability to amplify DNA in the polymerase chain reaction (PCR), thermostable DNA polymerases that were isolated from hyperthermophiles have been responsible for a significant breakthrough in the field of molecular biology. Taq polymerase, which originated from the bacterium *Thermus aquaticus*, is the best-known success story in this regard [38]. The enzymes that are produced by thermophiles have significant applications in the industries of detergent, food, feed, starch, textile, leather, pulp and paper, and pharmaceuticals [39, 40].

2.2 Halophilic microorganisms

Halophiles are organisms which have the ability to grow at higher salt concentration. They can survive in both moderate and extreme saline environment like salt mines and dead sea. Halophiles can thrive at 0.3–5.1 M NaCl concentrations [41]. Halophiles can also be discovered in ordinary habitats such as certain food products. For example, halophilic archaea are believed to play a key role in the fermentation process of kimchi, which is a popular dish in Korea [42]. Halophilic bacteria produce negatively charged enzymes which have promising applications in biotechnology due to their unique properties. The high concentration of halophiles and the lack of non-halophilic pollutants in hypersaline brines make them suited for a wide variety of biotechnological applications. In addition to their potential use in bioremediation and bio fermentation, halophiles may be a good source for obtaining a wide variety of other novel biomolecules, such as stable enzymes and biopolymers. Different studies are conducted on the conversion of plant and animal polymers in high salt environments, including the production of biofuels. Cellulolytic activity was detected in *Haloarcula* isolated from a Turkish salt mine at high salinities [43]. A strain from a Chinese salt lake showed cellulolytic activity in non-polar solvents, and it can ferment bioethanol from alkali-pretreated rice straw [44]. A species of *Halolactibacillus* was found able convert raw maize starch into compounds that could be then processed into bioethanol [45]. It was discovered that an extremely halophilic green alga known as *Dunaliella salina* produces a high lipid content that is ideal for the production of biodiesel [46]. Other halophilic lipases have also been reported, some of which have potential in biodiesel production [47, 48].

2.3 Psychrophilic microorganisms

Psychrotolerant or psychrophilic are organisms that can thrive at temperatures of 15°C or lower. These microorganisms can live in a variety of cold habitats on earth, such as the polar regions, glaciers, ocean depths, shallow underground regions, upper atmosphere and refrigerated equipment [49]. The majority of these microorganisms are classified as members of the bacterial family such as *Pseudoalteromonas*, *Vibrio, Pseudomonas, Arthrobacter*, and *Bacillus*, [50], *Methanogenium, Halorubrum, penicillium,* an *Cladosporium* are examples of archea, fungi, and yeast [51]. Psychrophilic enzymes are those that are produced by microorganisms that are adapted to colder environments and having high catalytic efficiency at low temperatures. These characteristics offers a significant opportunity in the industries of detergent, textile, food, pharmaceutical, leather, brewing and wine, and paper and pulp. Psychrophiles and their enzymes have been suggested as an alternative in bioremediation of polluted soils and waste water [52]. The structural properties of cold-active enzymes, such as a decrease in core hydrophobicity, a decrease in ionic interactions, an increase in surface charge, and longer surface loops, contribute to the flexibility required for optimal activity at low temperatures [53].

2.4 Alkaliphilic and acidophilic microorganisms

2.4.1 Alkaliphiles

Microorganisms which can withstand extreme alkaline conditions i.e., pH value greater than 9. One of the most common examples of naturally occurring alkaline environments are the soda lakes (with a pH between 10 and 12) or extremely saline [54]. Even though these conditions are extremely harsh, they can be home to a wide variety of bacteria and archaea and are among the most metabolically active marine ecosystems due to the presence of alkaphilic cyanobacteria [55].

The alkaliphiles offer a significant amount of potential for use in biotechnology. Alkaliphiles have the potential of producing different industrially important enzymes like proteases, amylases, cellulases, lipases, xylanases, pullulanases, pectinases, and chitinase [56]. The most important uses of these enzymes are in the production of detergents, the dehairing of hides, the production of pulp and paper, the hydrolysis of starch, and the preparation of food [57].

2.4.2 Acidophiles

Acidophiles are organisms which can withstand extreme lower pH, like hot springs and mine drainage systems. Acidophiles can be found in both natural and man-made environments. Acidophiles and thermophiles are frequently grouped together because thermophilic conditions are present in the majority of acidophilic habitats [58]. Bioleaching is the main application of acidophiles through which microorganisms decompose metal ores in order to remove the metal ions into solution. These metal ions may be then harvested, which enables economical metal extraction from low-grade ores. Among the most valuable properties of acidophiles is their ability to bioleach a variety of metal ions, including heavy metals, which are generally harmful to other organisms [59]. Acidophiles also have the ability to break down other hazardous organic molecules (such aliphatic compounds), and can be used to bioremediate acid mine drainage systems [60]. Enzymes isolated from acidophiles can be used in different industries like detergents and leather (**Table 1**).

2.5 Piezophilic microorganisms

Piezophiles and barophiles refer to organisms that are able to adapt to high levels of barometric pressure. Several different piezophiles have been successfully cultivated, however this process require the utilization of complex apparatus to maintain necessary pressures, such as hydraulic pumps and complex gas systems, which are able to keep up pressures up to 38 MPa [61]. Since many laboratories lack the resources and expertise to obtain and safely operate the specialized equipment required to culture these organisms, an increasing number of researches are relying on non-culturing methods, such as genomics, to learn more about them. Hydrothermal vents in the deep sea are well-studied examples of a high-pressure niche. Hydrothermal vents in the deep ocean are a prominent example of a high-pressure niche that has been extensively researched [62]. Piezophilic enzymes have the ability to withstand extreme pressure without requiring any pressure-related modifications. Enzymes extracted from piezophiles are resistant to high pressure and do not require any specialized modifications related to pressure. Enzymes isolated from piezophilic microbes have great potential in biotechnology (**Table 1**), especially in the food industry where high pressure is used to process and sterilize food [63]. Another enzymes chymotrypsin isolated from piezophilic microorganisms can function at high pressure and temperature and is used in many industrial processes [64].

2.6 Radiophilic microorganisms

Radiation-resistant or radioresistant extremophiles are microorganisms that can survive at extremely high levels of radiation. They have been discovered at high UVR altitudes (mountain ranges) and in broad fields. Continuous ozone depletion had a significant impact on global biosphere exposure to ultraviolet radiation. In addition, radioactive wastes have been released into the environment due to the widespread utilization radioactive compounds and elements for energy, medicine, research, and in industries [65]. Nuclear disasters like Fukushima Daiichi in 2011 and Chernobyl in 1986 have increased radionuclides and radioisotopes in the environment. X-rays and Gamma radiation are two more kinds of environmental radiation that can cause harm to humans. Different kinds of microbes have found strategies to survive in high radiation environments despite the damaging effects of radiation on humans. The bacterium *Deinococcus radiodurans* is resistant to extremely high level of radiation, both ionizing and ultraviolet (> 1000 J/m^2) [66]. A number of bacteria, including *Rhodanobacter sp*. and *Desulfuromonas ferrireducens*, have been found to thrive in environments with elevated radioactive concentrations [67]. Researchers have found a correlation between the DNA repair mechanisms and the production of protective primary and secondary metabolic products of radioresistant organisms and their ability to withstand high dose of radiation [68]. Biotechnological techniques can stimulate or trigger the production of radiation-responsive metabolites, pigments, and enzymes, which can be used to create pharmaceuticals, particularly anticancer treatments, antibiotics, and commercially important agricultural products [69].

3. Limits of life and extremophiles

Searching for life beyond earth is linked to our understanding of life on earth. In order to detect possible extraterrestrial habitats, it is essential to know the conditions that can sustain earth life. This does not mean that other planets and moons cannot support life similar to that on earth. Although different types of life may have different origins and biochemistry, it is possible that studying life on earth can help us understand life elsewhere. It is also possible for life to exist in environments completely different from those on earth (for example, the bacterium *Deinococcus radiodurans* has been shown to withstand radiation levels far exceeding our natural environment, and *Escherichia coli* has been shown to be able to withstand pressures ten times higher than those found in the deepest ocean trenches) [70, 71]. Life as we know it depends on water, light source or chemical energy, nutrients like nitrogen, phosphorus, sulfur, iron, are just few of the 70 elements on earth that either need or interact with life [72]. Currently, the search for extraterrestrial life focuses on planets and moons that had liquid water, geological and geophysical factors that encourage the synthesis and polymerization of organic molecules, as well as energy sources and nutrients necessary to sustain life on these planets. Once life start on a planet, then evolution will work to fill every possible niche, even if some niches have environments substantially different from where life originated. Since our understanding of life is based on what we can observe and measure, it can only be applied to earth-based life. Based on the universal laws of chemistry and physics, we can extrapolate the conditions for life on other planets. This suggests that life requires a solvent, an energy source, and building blocks in order to survive [73].

Considering the fact that everything requires energy for their chemical reactions, redox chemistry seems to be universal. The emergence, evolution, and diversity of life has often been influenced by physicochemical gradients that create non-equilibrium redox conditions [74]. A proton gradient and redox gradient were likely the two main mechanisms involved in the origin of life, driving metabolism and growth [75]. As a result, the search for life's limits has expanded beyond temperature, pH, pressure, salinity, and radiation gradients to include energy and nutrient limits which can be considered as well [76, 77]. Temperature, pH, pressure, salinity, and radiation are all related and can affect nutrient and energy availability. There are some parameters that affect microbial diversity more than others, like temperature in geothermal waters [78], pH in soil communities [79], salinity in saline lakes [80], and water content in dry climates [11].

Some organisms can thrive under conditions that limit their growth or prove lethal to other organisms. Majority of earth's organisms are killed by extreme temperatures, pH levels, salt concentrations, toxic metals and radiation levels. However, organisms in all three domains have adapted to many extremes on earth [5]. As a result of the emergence of life on earth, microbes have colonized habitats encompassing almost every conceivable physicochemical factor. It was previously believed that terrestrial environments are not suitable for growth because of high temperatures and low water activity (desiccation). It is now known that environments with $MgCl_2$ above 2.3 M also inhibit life, and this is due to $MgCl_2$ denaturing macromolecules in biological systems [81]. These conditions aren't necessarily sterile; many organisms have adapted mechanisms to survive at temperatures above 100°C, or even in a desiccated environment. Very few environments that claim to be sterile are genuinely free of all life forms. The Atacama Desert in Chile has been found to contain a small amount of viable microbes despite it being one of the driest environments on earth and believed to be similar to the environment on Mars [82]. Some liquid water environments do not support life, like the high-brine liquid in sea-ice inclusions at −30°C and the water above 40°C in submarine hydrothermal vents [83]. However, microbes have been observed to be able to survive in many extreme conditions outside their normal growth range, which still indicates their survival ability [84].

The limits of life, extremophile characteristics, and astrobiology implications are discussed in different studies [7]. Many discussions about the limits of life concentrate on extremes of one physical or chemical condition, such as temperature, salinity, heavy metal concentrations, desiccation, and pH [85]. To survive in nutrient-poor conditions, organisms have evolved a variety of metabolic and physiological strategies. Researchers found that *Pelagibacter ubique* a cosmopolitan microorganism in oligotrophic oceans, grew only at in situ micromolar concentrations of organic carbon. Despite its small genome, *P. ubique* has all the genes it needs to grow independently (without help from other organisms) [86]. The lowest concentrations of organic substances that can support heterotroph growth are set by *P. ubique* and related marine oligotrophs. *P. ubique* and similar marine oligotrophs could be used as models to develop detection strategies for organisms in the Lake Vostok, the subsurface of Mars, and Europa's ocean [87].

4. Survival mechanism of extremophiles under harsh conditions

Extremophiles have adapted a wide variety of strategies to survive in the inhospitable environments (**Figure 1**). Proteins are easily denatured and unfolded when exposed to temperatures over their normal range, as this breaks down intracellular bonds which is harmful to that particular organism. To prevent protein from degradation due to high temperature, thermophilic microbes produce chaperones or thermosomes, which allow them to recover their protein structure and function even in harsh environments [88]. Bacteria that thrive in high temperatures have evolved hydrogen bonds that interact with hydrophobicity to prevent protein unfolding. In contrast, the enzymes of thermophilic bacteria are structurally stable due to the abundance of salt and disulfide bridges. Additionally, structural compactness, oligomerization, glycosylation, and hydrophobic interactions between subunits all contribute to thermo-resistance and are therefore essential for stability [89]. The chaperones DnaK, GroEL, and GroES also assist protein folding in thermophiles through the role of heat shock proteins (HSPs). The DNA-repair system responds to DNA damage as well (**Figure 1**). Thermophiles use branched chain fatty acids and polyamines (such as spermidine) to stabilize their membranes [90]. In order to resist UV stress, ultraviolet resistant extremophiles have developed a variety of strategies (**Figure 1**). DNA repair, chaperone induction, and active defense against UV-induced oxidative stress (e.g., glutathione accumulation) are involved in these strategies [91]. Radiotolerance has been associated with the ability of these microorganisms to repair DNA damage because they accumulate a high level of intracellular manganese and a low level of iron, conferring UV resistance to them [92].

The cellular cold-adaptability mechanisms (**Figure 1**) of psychrophiles allow them to survive in extremely cold environments [93]. The presence of unsaturated fatty acids, cyclopropane-containing fatty acids, and short chain fatty acids in membranes prevents membrane fluidity loss [94]. Another mechanism is the high synthesis of cold-shock proteins (CSPs) and chaperones that protect RNA and protein synthesis [95]. The third mechanism involves the synthesis of antifreeze proteins (AFPs) that

DOI: http://dx.doi.org/10.5772/intechopen.110471

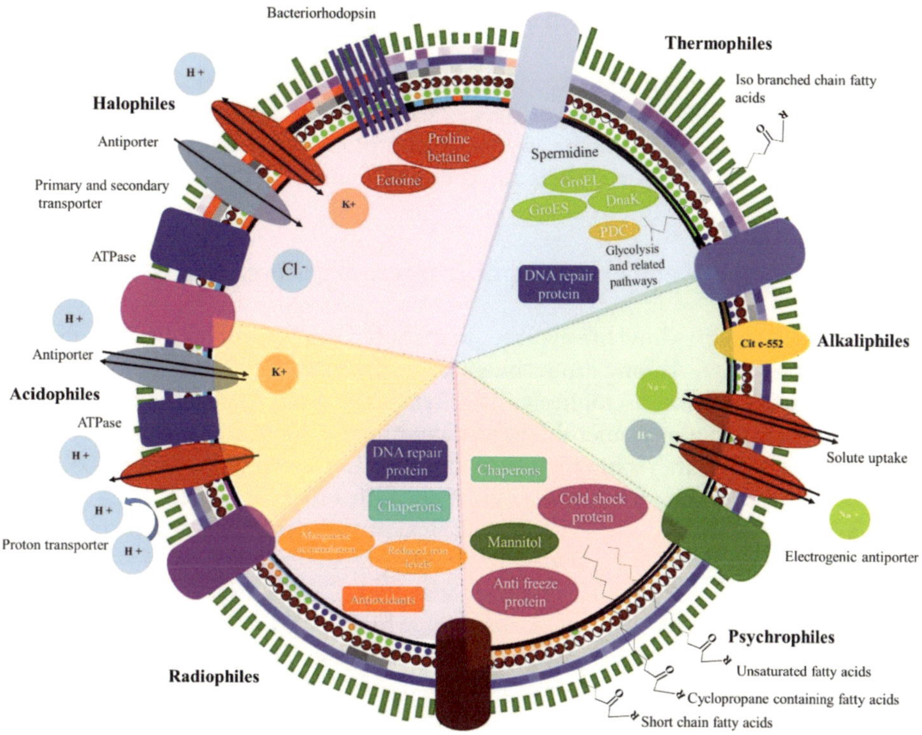

Figure 1.
Strategies adopted by different extremophiles for their survival in different extreme environments.

bind to ice crystals and cause thermal hysteresis [96]. A fourth mechanism is the accumulation of mannitol as a cryo-protectant [97].

In acidophiles, the cytoplasmic pH is maintained near neutrality in order to safeguard acid-labile cellular components. Adaptation to an acidic environment involves different mechanisms (**Figure 1**). In the first mechanism, protons are actively pumped to maintain a pH to 1 by proton flux systems. Several bacteria have been reported to efflux protons via transport pumps in the electron transport chain, as well as influx protons via F0F1-type ATP synthases [98]. Proton flux systems include primary proton pumps (symporters) and secondary proton pumps (e.g., cation/H+ antiporters). A second mechanism suppresses the entry of protons into the cytoplasm by lowering the permeability of the cell membrane. Inside the membrane, K+ ions form a positive potential that inhibits proton influx [99]. Thirdly, acidophiles possess a more advanced protein synthesis and DNA repair system than neutrophils. Acidophiles are induced by an external pH shift from 3.5 to 1.5 to produce proteins like chaperons involved in the heat shock response [100]. To survive in a saline environment, microorganisms have adapted different strategies (**Figure 1**). A first strategy involves maintaining a salt concentration within the cell equal to that in the environment. As a result, all intracellular systems have been adapted to the new environment. It is done by chloride and potassium transporters conjugated to bacteriorhodopsin and ATP synthase. Betaine and ectoine balance osmotic pressure by keeping intracellular salts low [101]. Another example of a microbial adaptation is the overexpression of a protein complex involved in DNA repair, replication, and recombination in two *Halobacterium sp. NRC-1* mutants that are capable of withstanding extremely high radiation levels

(LD50 > 11 kGy) [102]. To deal with high pH, alkaliphile bacteria use both symporters and antiporters (**Figure 1**). Symporters allow Na+ and other solutes into cells, and electrogenic antiporters produce a gradient of Na+ and H+ [103]. The respiratory system uses cytochrome C-552 to store electrons and hydrogen. By altering the distribution of ions (e.g., Na+), these systems allow protons and solutes to enter the cell, maintaining hydrosaline homeostasis and thermodynamic stability [104].

5. Relationship between extremophilic microorganism and astrobiology

Astrobiology studies how life evolved, distributed, and might continue in the universe in the future. Among other things, astrobiology brings a common biological perspective to astronomy, astrophysics, biochemistry, chemistry, extreme ecology, geology, molecular biology, microbiology, paleontology, physiology, planetary sciences, space exploration, technology, without omitting law and philosophy [105]. A major focus of astrobiology is finding evidence of life on other planets. This objective requires a clear understanding of the biophysical properties of life and the physical and chemical boundaries of earth's life [106]. Despite the fact that there are numerous definitions of life, none of them are widely accepted. Considering the gradual transition between abiotic structures and indisputable biological forms, any boundary between them must be based on a questionable standard. There is widespread agreement among biologists that the existence of DNA or RNA is a necessary condition for life to exist [107]. Mars and Jupiter's moon Europa are the best candidates for supporting life [108]. Several authors also proposed habitable environments on Venus and Saturn's moon Titan [105].

Different planetary bodies have similar environments like earth, having diverse range of each parameter. In order to discover habitable environments outside the earth, it is important to study extremophiles on earth to discover novel biosignatures. As far as biogeochemistry, nutrient composition, or topological similarities are concerned, extremophile habitats on earth share many similarities with those on other planetary bodies [109]. Based on this different (poly)extremophiles may persist in different planets depending on the planetary body. Halopsychrophiles may be able to survive on Titan, Ceres, and Europa because of their salty underground oceans [110], and on Mars because of its chlorine-rich brines [111, 112]. In addition, these living forms would have to withstand intense pressure, because the hydrostatic pressure in Titan's subterranean ocean varies between 140 and 800 MPa [113]. Despite the fact that these conditions are well outside the range of even the most extreme cultured piezophile (*Thermococcus piezophilus*, Pmax = 125 MPa) [114], microorganisms have successfully been exposed to pressures up to 2000 MPa and found to be metabolically active in fluid inclusions within type-IV ice [115]. Based on these observation it is possible that life may exist on other planets, such as Enceladus (Pmax = 50 MPa) and Europa (Pmax = 30 MPa) [5].

6. Conclusions

Extremophiles produce numerous extremozymes having different industrial applications, including agricultural, chemical, and medicinal. They are found in almost any habitat. Due to these advantages, extremozymes will be increasingly used in a wide range of consumer products. Extremophiles improve our understanding of macromolecular stability and physiochemical requirements for life. According to new research adaptations that enable survival under one stress, may also enable survival under other stress conditions. Finally, the study of extreme organisms contributes greatly to astrobiology's ongoing development. Understanding how life can thrive on earth may help astrobiologist better understand and locate potential life in other planetary bodies.

Author details

Nawab Ali[1*], Muhammad Nughman[1] and Syed Majid Shah[2]

1 Department of Biotechnology and Genetic Engineering, Kohat University of Science and Technology, Kohat, Khyber Pakhtunkhwa, Pakistan

2 Department of Pharmacy, Kohat University of Science and Technology, Kohat, Khyber Pakhtunkhwa, Pakistan

*Address all correspondence to: nawabali _ 1857@yahoo.com

References

[1] Horneck G, Klaus DM, Mancinelli RL. Space microbiology. Microbiology and Molecular Biology Reviews. 2010;**74**(1):121-156

[2] Yamagishi A et al. Environmental data and survival data of Deinococcus aetherius from the exposure facility of the Japan experimental module of the international space station obtained by the Tanpopo mission. Astrobiology. 2018;**18**(11):1369-1374

[3] Rampelotto PH. Extremophiles and extreme environments. Life. 2013;**3**(3):482-485

[4] Capece MC et al. Polyextremophiles and the constraints for terrestrial habitability. In: Polyextremophiles. Springer; 2013. pp. 3-59

[5] Ramirez J. Extremophiles: Defining the physical limits at which life can exist. Extremophiles. 2021

[6] MacElroy RD. Some comments on the evolution of extremophiles. Biosystems. 1974;**6**(1):74-75

[7] Rothschild LJ, Mancinelli RL. Life in extreme environments. Nature. 2001;**409**(6823):1092-1101

[8] Van Den Burg B. Extremophiles as a source for novel enzymes. Current Opinion in Microbiology. 2003;**6**(3):213-218

[9] Horikoshi K. Past, Present and Future of Extremophiles. Springer; 2008. pp. 1-2

[10] Seckbach J et al. Biodiversity and extremophiles. In: Life as We Know It. Springer; 2006. pp. 3-143

[11] Pikuta EV, Hoover RB, Tang J. Microbial extremophiles at the limits of life. Critical Reviews in Microbiology. 2007;**33**(3):183-209

[12] Bowers KJ, Mesbah NM, Wiegel J. Biodiversity of poly-extremophilic Bacteria: Does combining the extremes of high salt, alkaline pH and elevated temperature approach a physico-chemical boundary for life? Saline Systems. 2009;**5**(1):1-8

[13] Gupta G et al. Extremophiles: An overview of microorganism from extreme environment. International Journal of Agriculture, Environment and Biotechnology. 2014;**7**(2):371

[14] Dumorné K et al., Extremozymes: A Potential Source for Industrial Applications. 2017

[15] Shrestha N et al. Extremophiles for microbial-electrochemistry applications: A critical review. Bioresource Technology. 2018;**255**:318-330

[16] Zhu D et al. Recent development of extremophilic bacteria and their application in biorefinery. Frontiers in Bioengineering and Biotechnology. 2020;**2020**:483

[17] Sharma S, Kuila A, Sharma V. Enzymatic hydrolysis of thermochemically pretreated biomass using a mixture of cellulolytic enzymes produced from different fungal sources. Clean Technologies and Environmental Policy. 2017;**19**(5):1577-1584

[18] Mehta PK, Sehgal S. Microbial enzymes in food processing. In: Biocatalysis. Springer; 2019. pp. 255-275

[19] Lokko Y et al. Biotechnology and the bioeconomy—Towards inclusive and sustainable industrial development. New Biotechnology. 2018;**40**:5-10

[20] Ozdemir S et al. A novel raw starch hydrolyzing thermostable α-amylase

produced by newly isolated Bacillus mojavensis SO-10: Purification, characterization and usage in starch industries. Brazilian Archives of Biology and Technology. 2018;**61**

[21] Saiki RK et al. Primer-directed enzymatic amplification of DNA with a thermostable DNA polymerase. Science. 1988;**239**(4839):487-491

[22] Khusro A et al. Statistical optimization of thermo-alkali stable xylanase production from Bacillus tequilensis strain ARMATI. Electronic Journal of Biotechnology. 2016;**22**:16-25

[23] Mijts BN, Patel BK. Cloning, sequencing and expression of an α-amylase gene, amyA, from the thermophilic halophile Halothermothrix orenii and purification and biochemical characterization of the recombinant enzymeaaThe GenBank accession number for the sequence reported in this paper is AF442961. Microbiology. 2002;**148**(8):2343-2349

[24] Sanchez-Porro C et al. Screening and characterization of the protease CP1 produced by the moderately halophilic bacterium Pseudoalteromonas sp. strain CP76. Extremophiles. 2003;**7**(3):221-228

[25] Karasová-Lipovová P et al. The cloning, purification and characterisation of a cold-active β-galactosidase from the psychrotolerant Antarctic bacterium Arthrobacter sp. C2-2. Enzyme and Microbial Technology. 2003;**33**(6):836-844

[26] Zeng R et al. Cold-active serine alkaline protease from the psychrophilic bacterium Pseudomonas strain DY-A: Enzyme purification and characterization. Extremophiles. 2003;**7**(4):335-337

[27] Chang P et al. Cloning and characterization of two thermostable xylanases from an alkaliphilic Bacillus firmus. Biochemical and Biophysical Research Communications. 2004;**319**(3):1017-1025

[28] Rai SK, Konwarh R, Mukherjee AK. Purification, characterization and biotechnological application of an alkaline β-keratinase produced by Bacillus subtilis RM-01 in solid-state fermentation using chicken-feather as substrate. Biochemical Engineering Journal. 2009;**45**(3):218-225

[29] George SP, Ahmad A, Rao MB. Studies on carboxymethyl cellulase produced by an alkalothermophilic actinomycete. Bioresource Technology. 2001;**77**(2):171-175

[30] Deming JW et al. Isolation of an obligately barophilic bacterium and description of a new genus, Colwellia gen. nov. Systematic and Applied Microbiology. 1988;**10**(2):152-160

[31] Nogi Y, Masui N, Kato C. Photobacterium profundum sp. nov., a new, moderately barophilic bacterial species isolated from a deep-sea sediment. Extremophiles. 1998;**2**(1):1-8

[32] Llewellyn CA, Airs RL. Distribution and abundance of MAAs in 33 species of microalgae across 13 classes. Marine Drugs. 2010;**8**(4):1273-1291

[33] Singh B, Satyanarayana T. Production of phytate-hydrolyzing enzymes by thermophilic moulds. African Journal of Biotechnology. 2012;**11**(59):12314-12324

[34] Busk PK, Lange L. Cellulolytic potential of thermophilic species from four fungal orders. AMB Express. 2013;**3**(1):1-10

[35] Haki G, Rakshit S. Developments in industrially important thermostable enzymes: A review. Bioresource Technology. 2003;**89**(1):17-34

[36] Bhalla A et al. Improved lignocellulose conversion to biofuels with thermophilic bacteria and thermostable enzymes. Bioresource Technology. 2013;**128**:751-759

[37] Bertoldo C, Antranikian G. Starch-hydrolyzing enzymes from thermophilic archaea and bacteria. Current Opinion in Chemical Biology. 2002;**6**(2):151-160

[38] Chien A, Edgar DB, Trela JM. Deoxyribonucleic acid polymerase from the extreme thermophile Thermus aquaticus. Journal of Bacteriology. 1976;**127**(3):1550-1557

[39] Fujiwara S. Extremophiles: Developments of their special functions and potential resources. Journal of Bioscience and Bioengineering. 2002;**94**(6):518-525

[40] Kumar S, Nussinov R. How do thermophilic proteins deal with heat? Cellular and Molecular Life Sciences CMLS. 2001;**58**(9):1216-1233

[41] Oren A et al. Halobaculum gomorrense gen. nov., sp. nov., a novel extremely halophilic archaeon from the Dead Sea. International Journal of Systematic and Evolutionary Microbiology. 1995;**45**(4):747-754

[42] Chang H-W et al. Analysis of yeast and archaeal population dynamics in kimchi using denaturing gradient gel electrophoresis. International Journal of Food Microbiology. 2008;**126**(1-2):159-166

[43] Yildiz E, Ozcan B, Caliskan M. Isolation, characterization and phylogenetic analysis of halophilic archaea from a salt mine in Central Anatolia (Turkey). Polish Journal of Microbiology. 2012;**61**(2):111-117

[44] Yu HY, Li X. Characterization of an organic solvent-tolerant thermostable glucoamylase from a halophilic isolate, H alolactibacillus sp. SK 71 and its application in raw starch hydrolysis for bioethanol production. Biotechnology Progress. 2014;**30**(6):1262-1268

[45] Gao Y, Yang M, Wang C. Nutrient deprivation enhances lipid content in marine microalgae. Bioresource Technology. 2013;**147**:484-491

[46] Yilancioglu K et al. Oxidative stress is a mediator for increased lipid accumulation in a newly isolated Dunaliella salina strain. PLoS One. 2014;**9**(3):e91957

[47] Li X, Yu H-Y. Characterization of an organic solvent-tolerant lipase from Haloarcula sp. G41 and its application for biodiesel production. Folia Microbiologica. 2014;**59**(6):455-463

[48] Hou J et al. Characterization of genes for chitin catabolism in Haloferax mediterranei. Applied Microbiology and Biotechnology. 2014;**98**(3):1185-1194

[49] Cavicchioli R et al. Biotechnological uses of enzymes from psychrophiles. Microbial Biotechnology. 2011;**4**(4):449-460

[50] Collins G et al. Microbial community structure and methanogenic activity during start-up of psychrophilic anaerobic digesters treating synthetic industrial wastewaters. FEMS Microbiology Ecology. 2003;**46**(2):159-170

[51] Inagaki F et al. Biogeographical distribution and diversity of microbes in methane hydrate-bearing deep marine sediments on the Pacific Ocean Margin. Proceedings of the National Academy of Sciences. 2006;**103**(8):2815-2820

[52] Bhatia RK et al. Psychrophiles: A source of cold-adapted enzymes for energy

efficient biotechnological industrial processes. Journal of Environmental Chemical Engineering. 2021;**9**(1):104607

[53] Sarmiento F, Peralta R, Blamey JM. Cold and hot extremozymes: Industrial relevance and current trends. Frontiers in Bioengineering and Biotechnology. 2015;**3**:148

[54] Fujinami S, Fujisawa M. Industrial applications of alkaliphiles and their enzymes–Past, present and future. Environmental Technology. 2010;**31**(8-9):845-856

[55] Duckworth A et al. Phylogenetic diversity of soda lake alkaliphiles. FEMS Microbiology Ecology. 1996;**19**(3):181-191

[56] Wiegel J, Kevbrin V. Alkalither mophiles. Portland Press Ltd.; 2004

[57] Hasan F et al. Enzymes used in detergents: Lipases. African Journal of Biotechnology. 2010;**9**(31):4836-4844

[58] Elleuche S et al. Extremozymes—Biocatalysts with unique properties from extremophilic microorganisms. Current Opinion in Biotechnology. 2014;**29**:116-123

[59] Dopson M et al. Growth in sulfidic mineral environments: Metal resistance mechanisms in acidophilic micro-organisms. Microbiology. 2003;**149**(8):1959-1970

[60] Rani A, Souche YS, Goel R. Comparative assessment of in situ bioremediation potential of cadmium resistant acidophilic Pseudomonas putida 62BN and alkalophilic Pseudomonas monteilli 97AN strains on soybean. International Biodeterioration & Biodegradation. 2009;**63**(1):62-66

[61] Zhang Y et al. Current developments in marine microbiology: High-pressure biotechnology and the genetic engineering of piezophiles. Current Opinion in Biotechnology. 2015;**33**:157-164

[62] Jebbar M et al. Microbial diversity and adaptation to high hydrostatic pressure in deep-sea hydrothermal vents prokaryotes. Extremophiles. 2015;**19**(4):721-740

[63] Smedile F et al. Adaptations to high pressure of Nautilia sp. strain PV-1, a piezophilic campylobacterium (aka Epsilonproteobacterium) isolated from a deep-sea hydrothermal vent. Environmental Microbiology. 2022;**24**(12):6164-6183

[64] Leanwala AP. Application of Extremophiles in Food Industries, in Physiology, Genomics, and Biotechnological Applications of Extremophiles. IGI Global; 2022. pp. 251-259

[65] Gabani P, Singh OV. Radiation-resistant extremophiles and their potential in biotechnology and therapeutics. Applied Microbiology and Biotechnology. 2013;**97**(3):993-1004

[66] Yuan M et al. Deinococcus gobiensis sp. nov., an extremely radiation-resistant bacterium. International Journal of Systematic and Evolutionary Microbiology. 2009;**59**(6):1513-1517

[67] Green SJ et al. Denitrifying bacteria from the genus Rhodanobacter dominate bacterial communities in the highly contaminated subsurface of a nuclear legacy waste site. Applied and Environmental Microbiology. 2012;**78**(4):1039-1047

[68] Singh OV, Gabani P. Extremophiles: Radiation resistance microbial reserves and therapeutic implications. Journal of Applied Microbiology. 2011;**110**(4):851-861

[69] Murthy PS et al. Extremophiles as a source of biotechnological products, in Physiology, Genomics, and Biotechnological Applications of Extremophiles. IGI Global; 2022. pp. 308-333

[70] Cox MM, Battista JR. Deinococcus radiodurans—The consummate survivor. Nature Reviews Microbiology. 2005;**3**(11):882-892

[71] Sharma A et al. Microbial activity at gigapascal pressures. Science. 2002;**295**(5559):1514-1516

[72] Wackett LP, Dodge AG, Ellis LB. Microbial genomics and the periodic table. Applied and Environmental Microbiology. 2004;**70**(2):647-655

[73] Schwieterman EW et al. Exoplanet biosignatures: A review of remotely detectable signs of life. Astrobiology. 2018;**18**(6):663-708

[74] Covone G, Giovannelli D. Stellar Metallicity Is a Key Parameter for the Search of Life in the Universe. arXiv preprint arXiv:2207.03748. 2022

[75] Herschy B et al. An origin-of-life reactor to simulate alkaline hydrothermal vents. Journal of Molecular Evolution. 2014;**79**(5):213-227

[76] Hoehler TM, Jørgensen BB. Microbial life under extreme energy limitation. Nature Reviews Microbiology. 2013;**11**(2):83-94

[77] LaRowe DE, Amend JP. Power limits for microbial life. Frontiers in Microbiology. 2015;**6**:718

[78] Sharp CE et al. Distribution and diversity of Verrucomicrobia methanotrophs in geothermal and acidic environments. Environmental Microbiology. 2014;**16**(6):1867-1878

[79] Rousk J et al. Soil bacterial and fungal communities across a pH gradient in an arable soil. The ISME Journal. 2010;**4**(10):1340-1351

[80] Yang J et al. Salinity shapes microbial diversity and community structure in surface sediments of the Qinghai-Tibetan Lakes. Scientific Reports. 2016;**6**(1):1-6

[81] Hallsworth JE et al. Limits of life in MgCl2-containing environments: Chaotropicity defines the window. Environmental Microbiology. 2007;**9**(3):801-813

[82] Navarro-González R et al. Mars-like soils in the Atacama Desert, Chile, and the dry limit of microbial life. Science. 2003;**302**(5647):1018-1021

[83] Kelley DS, Baross JA, Delaney JR. Volcanoes, fluids, and life at mid-ocean ridge spreading centers. Annual Review of Earth and Planetary Sciences. 2002;**30**(1):385-491

[84] Baross JA, Deming JW. Growth at high temperatures: Isolation and taxonomy, physiology, and ecology. In: The Microbiology of Deep-sea Hydrothermal Vents. 1995. pp. 169-217

[85] Board SS, Council NR. The Limits of Organic Life in Planetary Systems. National Academies Press; 2007

[86] Giovannoni SJ et al. Genome streamlining in a cosmopolitan oceanic bacterium. Science. 2005;**309**(5738):1242-1245

[87] Belt G. Extremophiles, the physicochemical limits of life (growth and survival). In: Complete Course in Astrobiology. 2007

[88] Annamalai N, Rajeswari MV, Balasubramanian T. Thermostable and alkaline cellulases from marine sources.

In: New and Future Developments in Microbial Biotechnology and Bioengineering. Elsevier; 2016. pp. 91-98

[89] Chakraborty A et al. Metabolic responses of thermophilic endospores to sudden heat-induced perturbation in marine sediment samples. Frontiers in Microbiology. 2022;**13**

[90] Orellana R et al. Living at the frontiers of life: Extremophiles in Chile and their potential for bioremediation. Frontiers in Microbiology. 2018;**9**:2309

[91] Pérez V et al. Bacterial survival under extreme UV radiation: A comparative proteomics study of Rhodobacter sp., isolated from high altitude wetlands in Chile. Frontiers in Microbiology. 2017;**8**:1173

[92] Paulino-Lima IG et al. Extremely high UV-C radiation resistant microorganisms from desert environments with different manganese concentrations. Journal of Photochemistry and Photobiology B: Biology. 2016;**163**:327-336

[93] Jin S, Wang Y, Zhao X. Cold-adaptive mechanism of psychrophilic bacteria in food and its application. Microbial Pathogenesis. 2022:105652

[94] Bradburn MJ et al. Survival analysis part III: Multivariate data analysis–choosing a model and assessing its adequacy and fit. British Journal of Cancer. 2003;**89**(4):605-611

[95] Godin-Roulling A et al. Functional adaptations of the bacterial chaperone trigger factor to extreme environmental temperatures. Environmental Microbiology. 2015;**17**(7):2407-2420

[96] Muñoz PA et al. Structure and application of antifreeze proteins from Antarctic bacteria. Microbial Cell Factories. 2017;**16**:1-13

[97] Baraúna RA et al. A proteomic perspective on the bacterial adaptation to cold: Integrating OMICs data of the psychrotrophic bacterium Exiguobacterium antarcticum B7. Proteomes. 2017;**5**(1):9

[98] Michels M, Bakker EP. Generation of a large, protonophore-sensitive proton motive force and pH difference in the acidophilic bacteria Thermoplasma acidophilum and Bacillus acidocaldarius. Journal of Bacteriology. 1985;**161**(1):231-237

[99] Christel S et al. Multi-omics reveals the lifestyle of the acidophilic, mineral-oxidizing model species Leptospirillum ferriphilum T. Applied and Environmental Microbiology. 2018;**84**(3):e02091-e02017

[100] Amaro A et al. Effect of external pH perturbations on in vivo protein synthesis by the acidophilic bacterium thiobacillus ferrooxidans. Journal of Bacteriology. 1991;**173**(2):910-915

[101] Margesin R, Schinner F. Bioremediation (natural attenuation and biostimulation) of diesel-oil-contaminated soil in an alpine glacier skiing area. Applied and Environmental Microbiology. 2001;**67**(7):3127-3133

[102] Byrne RT et al. Evolution of extreme resistance to ionizing radiation via genetic adaptation of DNA repair. eLife. 2014;**3**:e01322

[103] Krulwich TA et al. pH homeostasis and ATP synthesis: Studies of two processes that necessitate inward proton translocation in extremely alkaliphilic Bacillus species. Extremophiles. 1998;**2**(3):217-222

[104] Matsuno T, Yumoto I. Bioenergetics and the role of soluble cytochromes c for alkaline adaptation in gram-negative

alkaliphilic Pseudomonas. BioMed Research International. 2015;**2015**

[105] Javaux EJ. Extreme life on earth—Past, present and possibly beyond. Research in Microbiology. 2006;**157**(1):37-48

[106] Trent JD. Extremophiles in astrobiology: Per Ardua ad Astra. Gravitational and Space Biology. 2000;**13**(2):5-12

[107] Van Loon A. The needless search for extraterrestrial fossils on earth. Earth-Science Reviews. 2005;**68**(3-4):335-346

[108] Seckbach J, Chela-Flores J. Frontiers of extremophilic microorganisms: from life on the edge to astrobiology. In: Exo-/ Astro-Biology. 2001

[109] Merino N et al. Living at the extremes: Extremophiles and the limits of life in a planetary context. Frontiers in Microbiology. 2019;**10**:780

[110] Cavicchioli R. Extremophiles and the search for extraterrestrial life. Astrobiology. 2002;**2**(3):281-292

[111] Clifford SM et al. Depth of the Martian cryosphere: Revised estimates and implications for the existence and detection of subpermafrost groundwater. Journal of Geophysical Research: Planets. 2010;**115**(E7)

[112] Jones EG, Lineweaver CH, Clarke JD. An extensive phase space for the potential martian biosphere. Astrobiology. 2011;**11**(10):1017-1033

[113] Stevenson A et al. Multiplication of microbes below 0.690 water activity: Implications for terrestrial and extraterrestrial life. Environmental Microbiology. 2015;**17**(2):257-277

[114] Dalmasso C et al. Thermococcus piezophilus sp. nov., a novel hyperthermophilic and piezophilic archaeon with a broad pressure range for growth, isolated from a deepest hydrothermal vent at the mid-Cayman rise. Systematic and Applied Microbiology. 2016;**39**(7):440-444

[115] Kish A et al. High-pressure tolerance in Halobacterium salinarum NRC-1 and other non-piezophilic prokaryotes. Extremophiles. 2012;**16**(2):355-361

Chapter 6

Genome Analysis Provides Insights into the Osmoadaptation Mechanisms of *Halomonas titanicae*

Afef Najjari

Abstract

Here, we report the osmoadaptation strategies adopted by the halotolerant species *Halomonas titanicae* BH1(T) inferred from genome sequence analysis. BH strain was isolated in 2010 from a rusticated sample collected in 1991 from the wreck of the Titanic, genome deposited in the database under the accession number (CP059082.1). It showed a high salt tolerance ranging from 0.5 to 25% NaCl (w/v) (optimal growth at 10% NaCl) with no growth in the absence of NaCl. The phylogenomic analysis showed that the BH1 strain is more closely related to the *Halomonas sedementi* QX-2, a strain isolated from deep-sea sediments. The RAST (Rapid Annotation using Subsystem Technology) annotation revealed divergent mechanisms involved in the primary and secondary response to osmotic stress citing protein implicated in potassium transport, periplasmic glucan synthesis, choline and betaine upake system, biosynthesis of glycine-betaine, ectoine, and proline. These findings provide an overview of the osmoadaptive mechanisms of *H. titanicae* BH1, and could offer helpful information to future biotechnological applications like osmolyte synthesis and related applications.

Keywords: *Halomonas titanicae* BH1, osmoadaptation, osmolytes, genome sequence

1. Introduction

Halomonas titanicae BH1 type strain was isolated from rusticlesis collected from the RMS Titanic wreck site [1]. It is a Gram-negative, heterotrophic, aerobic rod, and motile bacterium. It belongs to *Halomonadaceae* family [2]. Currently, there are 119 *Halomonas* species (https://lpsn.dsmz.de/) and interestingly, most of them are isolated from marine or hypersaline habitats [3]. They are slightly to moderately halophilic and oligotrophic organisms [4]. *H. titanicae* BH1 strain showed an ability to grow in media with 0.5–25% NaCl with no growth in the absence of NaCl [1], which reflects an ability to tolerate osmotic stress fluctuations. Generally, to cope with osmotic stress halophilic microorganisms deploy multiple strategies to regulate their internal osmotic pressure through the accumulation of organic or inorganic compatible solutes [5–7] citing (i) The primary response to osmotic stress: the early response consists of water influx either in or out of the cells through dedicated membrane channels called aquaporin. This mechanism is coupled with a salt-in strategy [5].

These changes are detected by osmosensors. It was suggested that these osmosensors are the aquaporins themselves and the stretch-activated/mechanosensitive channels (sensitive to cell turgor and intracellular tension) in eukaryotic cells [8]; (ii) the salt-in strategy: this strategy involves the accumulation of KCl. The exclusion of Na + from the cytoplasm is achieved through a Na+/H+ antiport (*Nha*C) located at the cytoplasmic membrane. Generally, K+ ions move in passively *via* a trkH system under the impulse of the membrane pressure. The main source of energy for the expulsion of Na+ and the accumulation of K+ in cells is the electrochemical potential difference of protons. This potential difference is due to both the transport of electrons in the respiratory chain, as well as the proton gradient formed during ATP synthesis by membrane ATPases and K+ antiporters (KdpABC). The influx of cations must be balanced by an equivalent number of anions. The movement of anions such as chloride is coupled to the energy of the membrane potential. It penetrates through a Na+/Cl– symport [5, 7]; (iii) The compatible osmolytes accumulation strategy: it employs the exclusion of intracellular salt ions simultaneously synthesizing or accumulating high concentrations of compatible solutes [5]. Compatible solutes can be sugars (sucrose, trehalose) and their derivatives (sulfotrehalose, glucosylglycerol), some amino acids and derivatives (proline, glutamic acid, glutamine, and glycine betaine), ectoine (and derivatives), or polyalcohols (glycerol, arabitol, and mannitol) [5]. These osmoprotectants can be synthesized or taken from the external environment without interfering with cellular metabolism. These solutes help to maintain turgor, pressure, cell volume, and electrolyte concentration. They can be released into the external environment either by other producing microorganisms following a drop in osmolarity or by decaying plant, animal, or bacterial cells [5, 7, 9, 10].

Here, we attempt to unravel the mechanisms of osmoadaption of adopted by *H. titanicae* BH1(T) strain to cope with the osmotic stress, based on *in silico* whole genome sequence analyses.

2. Material and method

2.1 *H. titanicae* BH1 genome sequence analysis

H. titanicae BH1 bacterium was isolated from a sample of rusticles collected from the RMS Titanic wreck site [1]. BH1 is Gram-negative cell, heterotrophic, aerobic rod, and motile by peritrichous flagella. Phylogenetically this organism belongs to the *Gammaproteobacteria* class within *Halomonadaceae* family [1]. The genome sequence of BH1 strain was retrieved from GenBank database under the accession number NZ_AOPO00000000 [11]. Cluster of Orthologous Groups (COG) functional categories was performed based on database of clusters of ortholgous genes.

2.2 Phylogenomic analysis of *H. titanicae* BH1

The whole-genome-based taxonomic analysis was performed by type strain genome server (TYGS) (at https://tygs.dsmz.de) [12]. The phylogenomic tree assessment was carried out using FastME algorithm [13] from the Genome BLAST Distance Phylogeny (GBDP). All pairwise-genome comparisons were conducted using Genome BLAST Distance Phylogeny approach (GBDP) under the algorithm "coverage" and distance formula d5 [14]. The trees were rooted at the midpoint [15]. Branch supports were inferred from 100 pseudo-bootstrap replicates.

2.3 Identification of genes involved in halotolerance in *H. titanicae* BH1

Genome sequence annotation was uploaded in Rapid Annotations using Subsystems Technology server (RAST, http://rast.nmpdr.org/) to identify potential genes and subsystems involved in the osomoadaptation pathways.

3. Results and discussions

3.1 Genome features and phylogenomic analysis

H. titanicae BH1 has been deposited in several culture collections as ATCC BAA-1257, CECT 7585, JCM 16411, and LMG 25388. It was able to grow in media with 0.5 to 25% NaCl and no growth occurs in the absence of NaCl [1]. The draft genome of *H. titanicae* BH1 (NZ_AOPO00000000.1) [11] includes 5,28 Mb with a G + C content of 54.6% and is composed of 4,759 putative protein-coding genes (**Figure 1**) [11].

Whole genome-based phylogenetic tree conducted by TYGS illustrated that strain BH1 forms a distinct clade with and forms a distinct cluster with *Halomonas sedimenti* QX-2'(NZ_JACCGK000000000.1) (**Figure 2**). *H. sedimenti* QX-2 is a halophilic gram-negative bacterium that was isolated from deep-water sediments in the south-west Indian Ocean at a depth of 2699 m [16].

The distribution of predicted genes based on RAST annotation among the subsystem database (**Figure 3**) revealed that subsystems with "Amino Acids and Derivatives (414)" and "Carbohydrates (281)" were the most represented subsystem features. In addition, the annotated subsystem features denoted 107 genes associated with "Stress response."

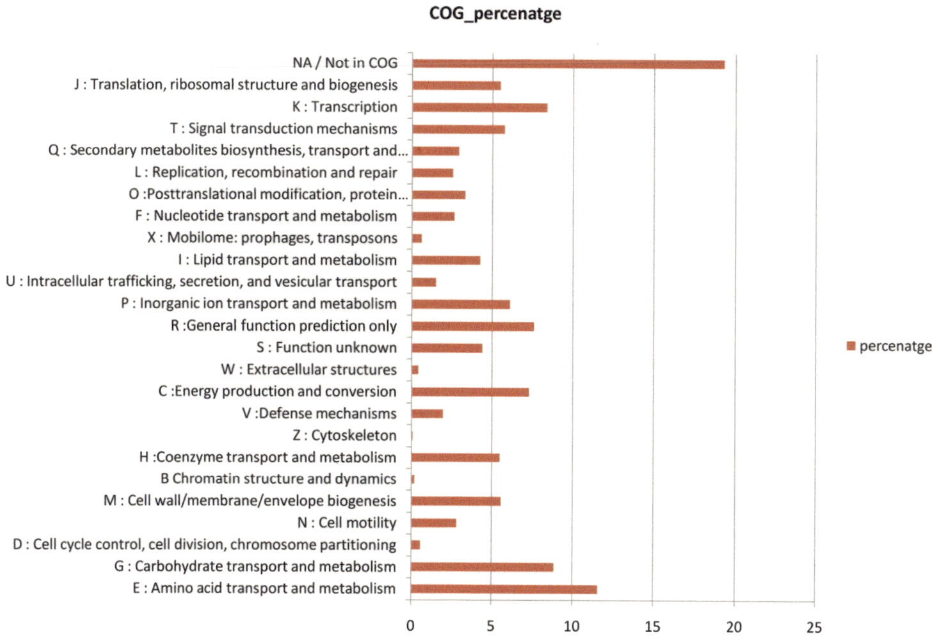

Figure 1.
The COG functional categories repartition of H. titanicaea BH1.

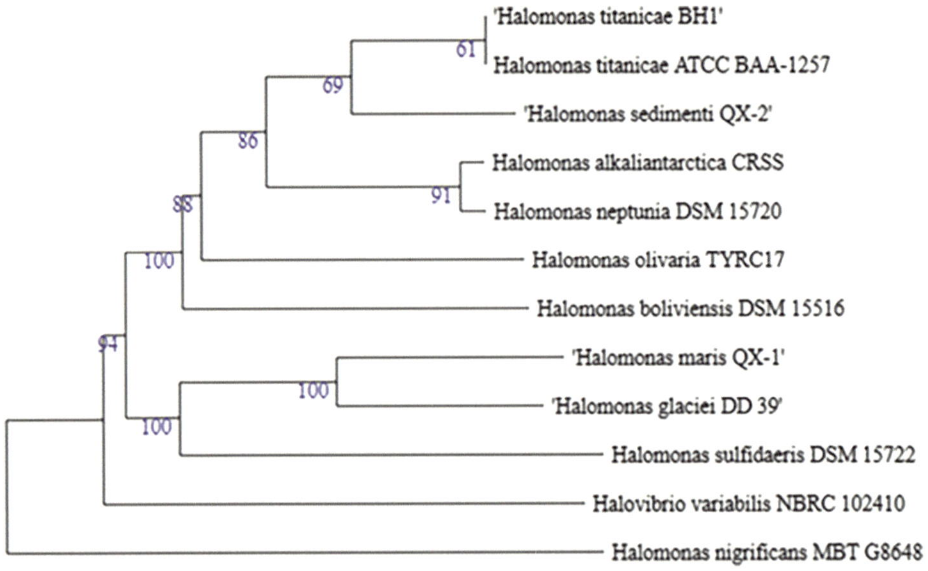

Figure 2.
Whole-genome-based phylogenetic tree highlighting the position of H. titanicaea strain BH1 to other closely related bacterial taxa. Trees are generated with fastme 2.1.6.1. The numbers above 60% from 100 replications. The tree was rooted at the midpoint.

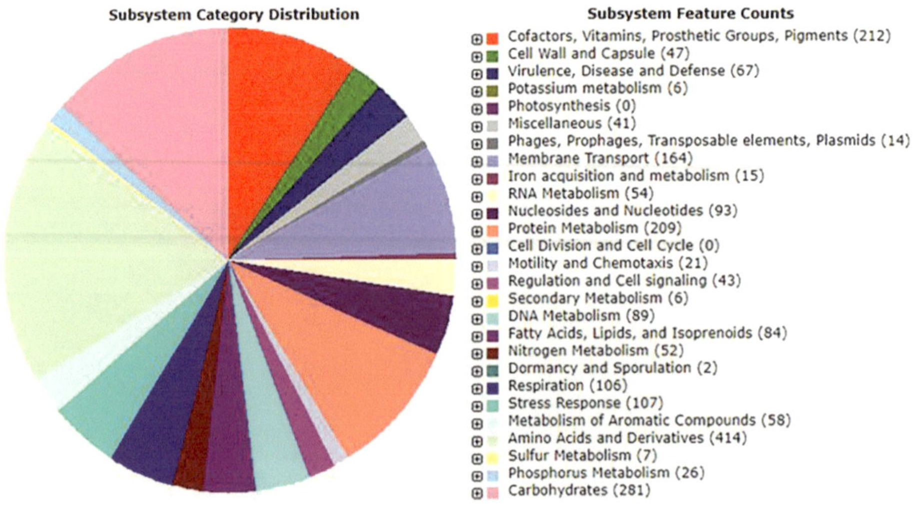

Figure 3.
Subsystem distribution of H. titanicaea strain BH1 genome based on RAST annotation server.

3.2 Genes related to osmoadaptation strategies in *H. titanicae strain* BH1

3.2.1 Primary responses to osmotic stress

3.2.1.1 Salt-in strategy: potassium transport

The Trk-type transporter system: Mining of the BH1 genome sequence showed three Trk-type transporters (**Table 1**) (i) TrkA system potassium uptake protein,

Subsystem	Gene	Predicted protein
Primary response to osmotic stress		
Trk-like transporter	*trkA*	Potassium transporter TrkA
	TrkI	Trk system potassium uptake protein
	TrkH	Trk system potassium uptake protein
kdpABC operon	*KdpE*	transcriptional regulatory protein
	KdpD	Histidine kinase
Synthesis of osmoregulated periplasmic glucans	*mdoH*	Glucan biosynthesis glucosyltransferase H
	mdoG	Glucan biosynthesis protein G precursor
Secondary response to osmotic stress—use of osmolytes		
Ectoine biosynthesis and uptake	*Asd*	L-aspartate-β-semialdehyde dehydrogenase
	Ask	L-aspartate kinase
	ectA	L-2,4-diaminobutyric acid acetyltransferase
	ectB	Diaminobutyrate-pyruvate aminotransferase
	ectC	L-ectoine synthase
	ectD	Ectoine hydroxylase
	TeaA	Ectoine-binding periplasmic protein
	TeaC	Ectoine TRAP transporter large permease protein
	TeaD	TRAP-T-associated universal stress protein
	TeaB	Ectoine TRAP transporter small permease protein
	eutD/doeB	Xaa-Pro dipeptidase
Choline and Betaine Uptake and Betaine Biosynthesis	*BetT2*	Choline transporter
	betI	
	BetL	Glycine betaine transporter
	betA	Choline dehydrogenase (EC 1.1.99.1)
	betB	betaine aldehyde dehydrogenase
	betC	Choline-sulfatase (EC 3.1.6.6)
	betI	HTH-type transcriptional regulator BetI
	DgcA	dimethylglycine demethylation protein
	Sarcosine oxydase	SoxBDAG
	OpuD	Glycine betaine transporter
	opuA	Glycine betaine ABC transport system, glycine betaine-binding protein
	proU	L-proline glycine betaine binding ABC transporter protein
	GbuC	Glycine betaine/carnitine transport binding protein
	ATPase	Quaternary-amine-transporting ATPase
	YehY	Glycine betaine uptake system permease protein
	YehW	Glycine betaine uptake system permease protein
	BCCT transporter	L-carnitine/gamma-butyrobetaine antiporter

Subsystem	Gene	Predicted protein
Proline synthesis and uptake	*proZ*	proline/glycine betaine ABC transporter permease
	proA	glutamate semialdehyde dehydrogenase
	proJ	the glutamate 5-kinase
	proH	Glutamate-5-semialdehyde dehydrogenase
	*pro*V	proline/glycine betaine ABC transporter permease
	Proline racemase	Proline, 4-hydroxyproline uptake and utilization
	PutR for proline utilization,	Proline, 4-hydroxyproline uptake and utilization

Table 1.
Genes and encoded proteins implicated in halotolerance in Halomonas titanicae BH1.

serves to control the activity of the potassium translocating subunit (ii) TrkH system potassium uptake protein, exhibits only a low affinity for K+ (iii) TrkI system potassium uptake protein, the main K+ transporter in osmotically adapted cells, exhibits medium affinity for K+. Several studies reported the role of K^+ uptake *via* Trk-like transporter in the primary response to osmotic stress in bacteria [17, 18].

Trk-type transporter systems have an important role in controlling the flux of potassium (K+) ions into cells. Generally, K+ ions enter passively *via* a uniport system (trkH) under the impulse of the membrane potential. The difference in potential is mainly due to the transport of electrons in the respiratory chain, as well as to the proton gradient generated during ATP synthesis *via* membrane ATPases and K+ antiport transporters (*Kdp*ABC). Here, BH1 genome sequence contains three genes (i) *Kdp*E: transcriptional regulatory protein involved in the regulation of the kdp operon; (ii) Histidine kinase KdpD, involved in the regulation of the kdp operon; and (iii) *Kdp*D may function as a membrane-associated protein kinase that phosphorylates KdpE in response to environmental signals [19].

3.2.1.2 Periplasmic glucans synthesis

Periplasmic glucans synthesis (OPG) in *H. titanicae* strain BH1 is carried out by the products of two genes, *mdoG* and *mdoH* (**Table 1**). MdoH catalyzes the production of linear β-1,2 polyglucose chains from the precursor UDP-glucose. MdoG function is unclear. Earlier studies showed that in some gram-negative bacteria, OPGs are synthesized at low osmolarity, which suggests that they play a part in the initial response to osmotic stress [20].

3.2.2 Secondary response to osmotic stress: biosynthesis and uptake of osmolytes

Osmolytes are organic molecules with low molecular weight, which can accumulate in cells at elevated concentrations without interfering with cell function, thanks to their high solubility and non-interaction with proteins [5, 7]. Analysis of the BH1 genome sequence revealed several osmolytes implicated in osmodapatation.

3.2.2.1 Ectoine biosynthesis and uptake

The complete biosynthetic pathway of ectoine/hydroxyectoine was identified in BH1 genome sequence (**Table 1, Figure 4**). In fact, the biosynthesis of ectoine

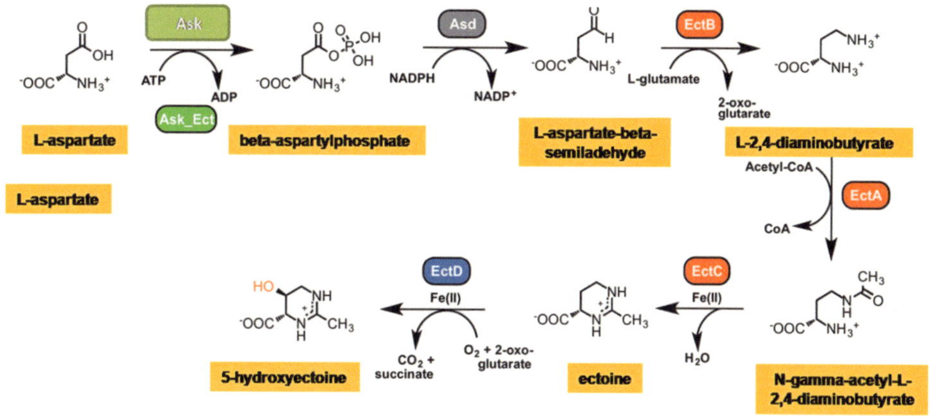

Figure 4.
Biosynthetic pathway for ectoines from aspartate in H. titanicaea strain BH1. The enzymes involved are Aspartate kinase (Ask), Aspartate semialdehyde dehydrogenase (Asd), L-diaminobutyric acid transaminase (EctB), L-diaminobutyric acid acetyl transferase (EctA), ectoine synthase (EctC), and ectoine hydroxylase (EctD).

from aspartate is catalyzed successively by five enzymes: L-aspartate kinase (Ask), L-aspartate-β-semialdehyde dehydrogenase (Asd), L-2,4-diaminobutyrate aminotransferase (EctB), L-2,4-diaminobutyrate acetyltransferase (EctA), and ectoine synthase (EctC). In addition, ectoine hydroxylase (EctD), which catalyzes the conversion of ectoine to hydroxyectoine, has been identified. The genes encoding EctB, EctA, and EctC are typically present as a gene cluster (ectABC) in some cases with ectD [21, 22].

In addition, the TeaABCD gene cluster, involved in the uptake of ectoine as a response to an osmotic shock, has also been identified (**Figure 4**, **Table 1**). *TeaABC*, is an osmoregulated transporter that catalyzes the uptake of ectoine and hydroxyectoine as a response to osmotic shock. TeaD (ATP-binding protein) negatively regulates the activity of the tripartite ATP-independent periplasmic ectoine transport system (TeaABC) [23].

3.2.2.2 Choline and glycine betaine uptake and biosynthesis

Osmotic adaptation can also be achieved by the uptake of betaine/choline osmolytes available in the environment. It should be noted that the uptake strategy is favored as it is significantly less energetically expensive than the *de novo* synthesis [7].

3.2.2.3 Uptake of choline and glycine betaine

Choline uptake is achieved by the BetT2 osmo-dependent choline transporter, transcribed *via* the betIBA operon (**Table 1**). For glycine uptake, several membrane transporters have been identified, notably (i) the glycine betaine transporter (BetL), a member of the BCCT family of transporters, displays a high affinity for glycine betaine uptake, (ii) OpuAB, a permease protein for the glycine betaine transport system. It is part of the binding protein-dependent transport family, (iii) glycine betaine transporter OpuD, is a single-component BCCT-type transporter, (iv) glycine betaine/carnitine transport binding protein GbuC, part of the ABC transporter complex GbuABC involved in glycine betaine and carnitine uptake, (v) quaternary-amine-transporting ATPase: involved in a multicomponent binding-protein-dependent transport system for glycine betaine, (vi) glycine betaine uptake system permease protein (YehY), part of ABC transporter complex involved in low-affinity glycine betaine uptake, (vii) glycine betaine uptake system permease protein (YehW), part of an ABC transporter complex involved in low-affinity glycine betaine uptake, and (viii) L-carnitine/gamma-butyrobetaine antiporter, which catalyzes the exchange of L-carnitine for gamma-butyrobetaine and related betaines, it belongs to the BCCT transporter family [24].

3.2.2.4 Glycine betaine biosynthesis

In silico whole genome analysis reveals that the biosynthesis of glycine betaine (GB) can be achieved *via* choline oxydation and then glycine demethylation

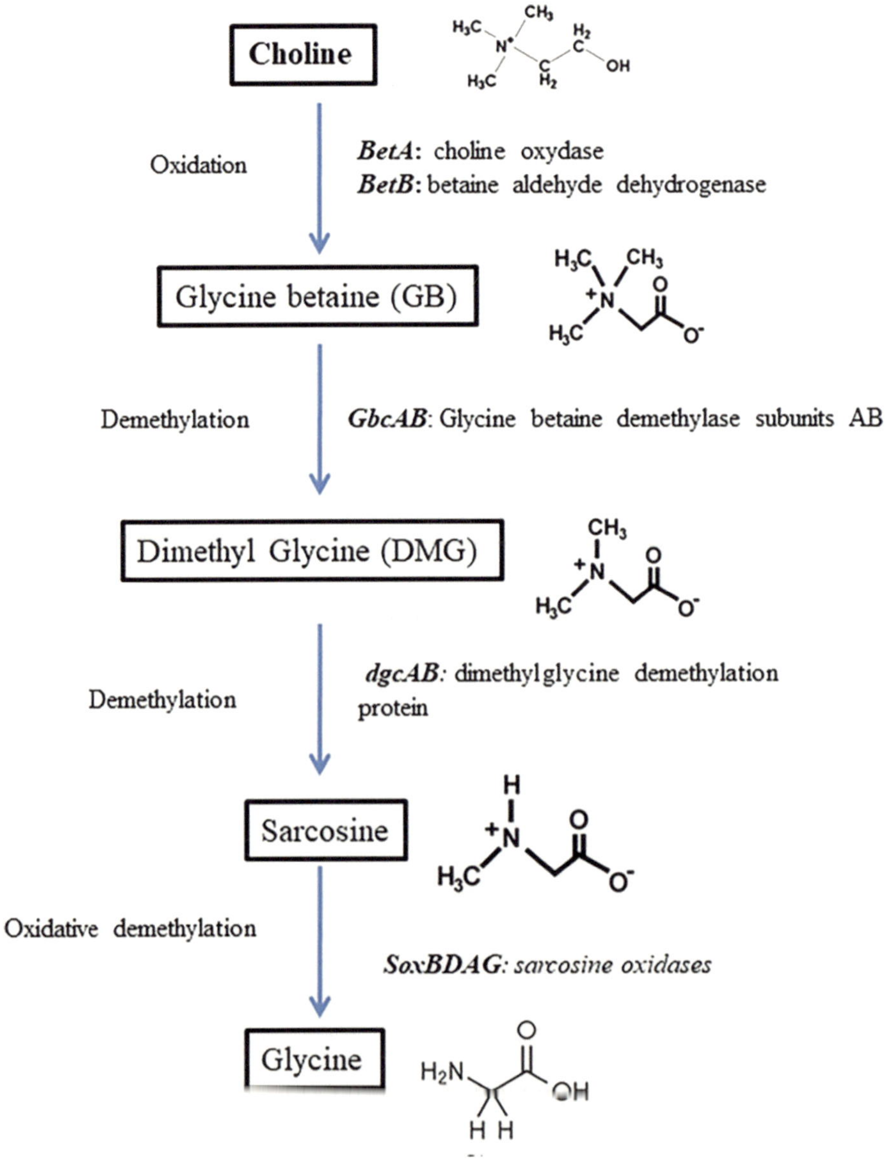

Figure 5.
Schematic representation of choline catabolic pathway of H. titanicae BH1.

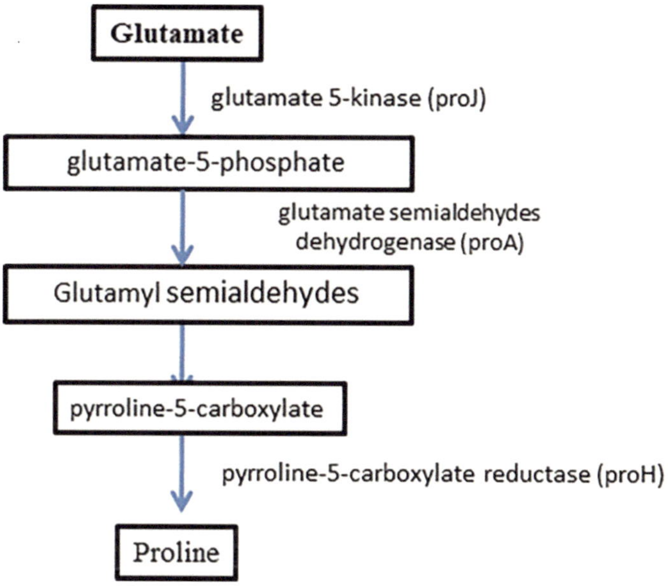

Figure 6.
Proline biosynthesis from glutamate in H. titanicae BH1.

(**Table 1**, **Figure 5**) [24]. The reaction is as follows: choline is first oxidized into GB by a pair of enzymes from the same operon, choline oxidase (BetA) and betaine aldehyde dehydrogenase (BetB). Then, the GB is demethylated *via* the oxygenase activity of GbcAB enzyme (Glycine betaine demethylase subunit A and B) to dimethylglycine (DMG). Next, the heterodimeric flavin-linked oxidoreductase DgcA (dimethylglycine demethylation protein) catalyzes the conversion of DMG to sarcosine. Finally, oxidative demethylation of sarcosine is conducted by a heterotetrameric enzyme (SoxBDAG) that generates glycine. Genes of this catabolic pathway have been recognized based on comparative genomic analysis. In fact, the same pathway was identified in the *Pseudomonas aeruginosa* species [25].

3.2.2.5 Proline biosynthesis and uptake

Glutamate is considered as the primary precursor amino acid for proline synthesis. The biosynthesis is catalyzed by three enzymes (**Table 1**, **Figure 6**) [26, 27]: First, glutamate 5-kinase (proJ), catalyzes the transfer of a phosphate group to glutamate to form L-glutamate 5-phosphate. Then, glutamate semialdehyde dehydrogenase (proA), catalyzes the NADPH-dependent reduction of L-glutamate 5-phosphate to L-glutamate 5-semialdehyde. Next, Pyrroline-5-carboxylate reductase (proH) catalyzes the reduction of 1-pyrroline-5-carboxylate (PCA) to L-proline. The uptake of proline is under osmotic control and is mediated by the osmoregulated glycine betaine transport systems and 4-hydroxyproline uptake and utilization (**Table 1**).

4. Conclusion

In conclusion, the analysis of the whole genome sequence gave us an insight into the osmotic adaptation mechanisms adopted by the *H. titanicae* BH1 strain face to osmotic stress. In fact, two main strategies were identified, the salt-in-cytoplasm and the solute accumulation or biosynthesis strategies. Genes implicated in several pathways were identified as well. This funding may facilitate an in-depth understanding of the transcription or the regulation of the metabolic pathways under stressful conditions.

Author details

Afef Najjari
Faculty of Sciences of Tunisia, University of Tunis El Manar, Tunis, Tunisia

*Address all correspondence to: afef.najjari@fst.utm.tn; najjariafef@gmail.com

References

[1] Sánchez-Porro C, Kaur B, Mann H, Ventosa A. *Halomonas titanicae* sp. nov., a halophilic bacterium isolated from the RMS titanic. International Journal of Systematic and Evolutionary Microbiology. 2010;**60**:2768-2774

[2] de la Haba RR, Sánchez-Porro C, Ventosa A. Taxonomy, phylogeny, and biotechnological interest of the family Halomonadaceae. In: Ventosa A, Oren A, Ma Y, editors. Halophiles and Hypersaline Environments. Heidelbeg, Germany: Springer; 2011. pp. 27-64

[3] Zhang T, Zhang X, Li Y, Yang N, Qiao L, Miao Z, et al. Study of osmoadaptation mechanisms of halophilic Halomonas alkaliphila XH26 under salt stress by transcriptome and ectoine analysis. Extremophiles. 2022;**26**(1):14

[4] Arahal DR, Ventosa A. The family *Halomonadaceae*. In: Dworkin M, Falkow S, Eosenberg E, Schleifer K-H, Stackebrandt E, editors. The Prokaryotes: A Handbook on the Biology of Bacteria. Singapore: Springer Science; 2006. pp. 811-835

[5] Roberts, m. f. Osmoadaptation and osmoregulation in archaea: Update 2004. Frontiers in Bioscience. 2004;**2004**:1999-2019

[6] Najjari A et al. Patterns and determinants of halophilic Archaea (class Halobacteria) diversity in Tunisian endorheic Salt Lakes and Sebkhet systems. Applied and Environmental Microbiology. 2015;**81**(13):4432-4441. DOI: 10.1128/aem.01097-15

[7] Youssef NH et al. Trehalose/2-sulfotrehalose biosynthesis and glycine-betaine uptake are widely spread mechanisms for osmoadaptation in the Halobacteriales. ISME Journal. 2014;**8**(3):636-649. DOI: 10.1038/ismej.2013.165

[8] Hill AE, Shachar-Hill Y. 'Are Aquaporins the Missing Transmembrane Osmosensors ?'. 2015. DOI: 10.1007/s00232-015-9790-0.

[9] Ferrer J, San E. Competition for water between protein (from Haloferax mediterranei) and cations Na + and K +: A quantum approach to problem. Theoretical Chemistry Accounts. 2016;**135**(9):1-8. DOI: 10.1007/s00214-016-1983-9

[10] Vaidya S, Dev K, Sourirajan A. Distinct Osmoadaptation strategies in the strict halophilic and halotolerant Bacteria isolated from Lunsu salt water body of north West Himalayas. Current Microbiology. 2018;**75**(7):888-895. DOI: 10.1007/s00284-018-1462-8

[11] Sánchez-Porro C, de la Haba RR, Cruz-Hernández N, González JM, Reyes-Guirao C, Navarro-Sampedro L, et al. Draft genome of the marine Gammaproteobacterium Halomonas titanicae. Genome Announcements. 2013;**1**(2):e0008313. DOI: 10.1128/genomeA.00083-13

[12] Meier-Kolthoff JP, Göker M. TYGS is an automated high-throughput platform for state-of-the-art genome-based taxonomy. Nature Communications. 2019;**10**:2182. DOI: 10.1038/s41467-019-10210-3

[13] Desper R, Gascuel O. Fast and accurate phylogeny reconstruction algorithms based on the minimum-evolution principle. Journal of Computational

Biology. 2002;**9**(5):687-705. DOI: 10.1089/10665270276103413б

[14] Meier-Kolthoff JP, Auch AF, Klenk H-P, Göker M. Genome sequence-based species delimitation with confidence intervals and improved distance functions. BMC Bioinformatics. 2013;**14**:60

[15] Farris JS. Estimating phylogenetic trees from distance matrices. The American Naturalist. 1972;**106**:645-668. DOI: 10.1086/282802

[16] Qiu X, Yu L, Cao X, Wu H, Xu G, Tang X. Halomonas sedimenti sp. nov., a halotolerant bacterium isolated from Deep-Sea sediment of the Southwest Indian Ocean. Current Microbiology. 2021 Apr;**78**(4):1662-1669. DOI: 10.1007/s00284-021-02425-9

[17] Bremer E, Krämer R. Responses of microorganisms to osmotic stress. Annual Review of Microbiology. 2019;**73**:313-334. DOI: 10.1146/annurev-micro-020518-115504

[18] Sleator RD, Hill C. Bacterial osmoadaptation: The role of osmolytes in bacterial stress and virulence. FEMS Microbiology Reviews. 2002;**26**:49-71. DOI: 10.1111/ j.1574-6976.2002.tb00598.x

[19] Price-Whelan A, Poon CK, Benson MA, Eidem TT, Roux CM, Boyd JM, et al. Transcriptional profiling of Staphylococcus aureus during growth in 2 M NaCl leads to clarification of physiological roles for Kdp and Ktr K+ uptake systems. mBio. 2013;**4**:e00407-e00413. DOI: 10.1128/mBio.00407- 13

[20] Kindzierski V, Raschke S, Knabe N, Siedler F, Scheffer B, Pflüger-Grau K, et al. Osmoregulation in the halophilic bacterium Halomonas elongata: A case study for integrative systems biology. PLoS One. 2017;**12**:e0168818. DOI: 10.1371/journal.pone.0168818

[21] Ono H, Sawada K, Khunajakr N, Tao T, Yamamoto M, Hiramoto M, et al. Characterization of biosynthetic enzymes for ectoine as a compatible solute in a moderately halophilic eubacterium, *Halomonas elongata*. Journal of Bacteriology. 1999;**181**:91-99

[22] Liu M, Liu H, Shi M, et al. Microbial production of ectoine and hydroxyectoine as high-value chemicals. Microbial Cell Factories. 2021;**20**:76. DOI: 10.1186/s12934-021-01567-6

[23] Tetsch L, Kunte HJ. The substrate-binding protein TeaA of the osmoregulated ectoine transporter TeaABC from Halomonas elongata: Purification and characterization of recombinant TeaA. FEMS Microbiology Letters. 2002;**211**(2):213-218. DOI: 10.1111/j.1574-6968.2002.tb11227.x

[24] Imhoff JF, Rahn T, Künzel S, Keller A, Neulinger SC. Osmotic adaptation and compatible solute biosynthesis of phototrophic Bacteria as revealed from genome analyses. Microorganisms. 2020;**9**(1):46

[25] Wargo MJ. Homeostasis and catabolism of choline and glycine betaine: Lessons from *Pseudomonas aeruginosa*. Applied Environmental Microbiology;**79**(7):2112-2120. DOI: 10.1128/AEM.03565-12

[26] Gowrishankar J. Identification of osmoresponsive genes in *Escherichia coli*: Evidence for participation of potassium and proline transport systems in osmoregulation. Journal of Bacteriology. 1985;**164**:434-445

[27] Saum SH, Müller V. Salinity-dependent switching of osmolyte strategies in a moderately halophilic bacterium: Glutamate induces proline biosynthesis in Halobacillus halophilus. Journal of Bacteriology. 2007;**189**:6968-6975. DOI: 10.1128/jb.00775-777